Adapting to an Uncertain Climate

Tiago Capela Lourenço • Ana Rovisco
Annemarie Groot • Carin Nilsson
Hans-Martin Füssel • Leendert van Bree
Roger B. Street
Editors

Adapting to an Uncertain Climate

Lessons From Practice

Editors
Tiago Capela Lourenço
Ana Rovisco
Faculty of Sciences
CCIAM (Centre for Climate Change,
  Impacts, Adaptation and Modelling)
University of Lisbon, Lisbon, Portugal

Carin Nilsson
Centre for Environmental
  and Climate Research
Lund University
Lund, Sweden

Leendert van Bree
Department of Spatial Planning
  and Quality of Living
PBL Netherlands Environmental
  Assessment Agency
The Hague, The Netherlands

Annemarie Groot
Alterra – Climate Change and Adaptive
  Land and Water Management
Wageningen University and Research Centre
Wageningen, Gelderland
The Netherlands

Hans-Martin Füssel
Air and Climate Change Programme
European Environment Agency
Copenhagen K, Denmark

Roger B. Street
UKCIP
Environmental Change Institute
University of Oxford
Oxford, UK

Chapter 3: Hans-Martin Füssel, (How Is Uncertainty Addressed in the Knowledge Base for National Adaptation Planning?)
© European Environment Agency, Copenhagen 2014

All rights reserved
No part of chapter 3 may be reproduced in any form or by any means electronic or mechanical, including photocopying, recording or by any information storage retrieval system, without a prior permission in writing. For permission, translation or reproduction rights please contact EEA (copyrights@eea.europa.eu)

ISBN 978-3-319-04875-8   ISBN 978-3-319-04876-5 (eBook)
DOI 10.1007/978-3-319-04876-5
Springer Cham Heidelberg New York Dordrecht London

Library of Congress Control Number: 2014936552

© Springer International Publishing Switzerland 2014
This work is subject to copyright. All rights are reserved by the Publisher, whether the whole or part of the material is concerned, specifically the rights of translation, reprinting, reuse of illustrations, recitation, broadcasting, reproduction on microfilms or in any other physical way, and transmission or information storage and retrieval, electronic adaptation, computer software, or by similar or dissimilar methodology now known or hereafter developed. Exempted from this legal reservation are brief excerpts in connection with reviews or scholarly analysis or material supplied specifically for the purpose of being entered and executed on a computer system, for exclusive use by the purchaser of the work. Duplication of this publication or parts thereof is permitted only under the provisions of the Copyright Law of the Publisher's location, in its current version, and permission for use must always be obtained from Springer. Permissions for use may be obtained through RightsLink at the Copyright Clearance Center. Violations are liable to prosecution under the respective Copyright Law.

The use of general descriptive names, registered names, trademarks, service marks, etc. in this publication does not imply, even in the absence of a specific statement, that such names are exempt from the relevant protective laws and regulations and therefore free for general use.

While the advice and information in this book are believed to be true and accurate at the date of publication, neither the authors nor the editors nor the publisher can accept any legal responsibility for any errors or omissions that may be made. The publisher makes no warranty, express or implied, with respect to the material contained herein.

Cover image caption: Praia de Coimbra 2011 by Hugo Costa

Printed on acid-free paper

Springer is part of Springer Science+Business Media (www.springer.com)

Sponsor

Contributors

Supported by

Foreword

According to the most recent report by the Intergovernmental Panel on Climate Change (IPCC), the warming of the climate system due to human activities is unequivocal. I cannot think of a better statement – may it seem contradictory – to introduce a book dealing with climate uncertainty.

Doubt, uncertainty, and indeed skepticism are inherent to science. Climate action, maybe beyond any other policy process, has been driven by climate science since the very recognition of the problem of climate change some few decades ago. Paradoxically, the inherent uncertainty of climate science was used by so-called climate skeptics to disregard climate action and the whole issue of climate change. Yet the scientific community, including through the IPCC, has kept providing ever-increasing data, analysis, and evidence from a multidisciplinary wealth of information, demonstrating beyond reasonable doubt that the planet is warming due to the buildup of greenhouse gases in the atmosphere.

Of course, uncertainty remains present in most of our political, economic, and social decisions, including those related to the changing climate. But whereas in other policy areas action would not usually be hindered on arguments of lack of absolute certainty in presence of highly likely facts, climate action has always been questioned, including through the use of fake arguments that wrongly mix up rigor, uncertainty, and likelihood. In the meantime, global warming continues, the increased impacts of both slow-onset events and of altered regimes of extreme weather events are a reality, and global sustainability keeps a distant goal for humankind.

The publication of this book is very welcome in this context and at this stage. It deals effectively with climate uncertainty, one of the most prominent and proclaimed barriers to developing effective adaptation action. It is true that climate policy needs to be tackled under significant uncertainty from several sources, as identified in Chap. 2. But on the other hand there are several options allowing us to start, such as no-regret, win-win, and cost-effective adaptation measures, particularly those useful to deal with on-going climate effects, or that will be needed in any case to help solve other problems. Further, there is a rapidly increasing knowledge base which is expected to reduce significantly uncertainty about changing climate patterns. Yet, some level of uncertainty will always remain, irrespective of scientific progress, as it is inherent to complex systems, to scientific methods, and indeed to reality as it is.

Climate change highlights the need to face shifting, albeit permanent, levels of uncertainty, setting a challenge for policy making with a long-term perspective. It calls for the development of flexible approaches, innovative governance, and other elements that might contribute to effective decision-making. Exploring these new approaches is one of the challenges for climate adaptation policy. The European Union, arguably a frontrunner in climate action, is already progressing in the development of approaches to adaptation that face the uncertainty challenge. Chapter 3 adequately shows how the EU and its Member States are dealing with this in the development of their adaptation strategies.

Remarkably, beyond the level of strategic planning, a body of knowledge on how to deal with uncertainty in adaptation for sectoral policies and local settings has started to emerge. Stakeholders from many areas are struggling to develop planning approaches that effectively deal with uncertainty and integrate it into medium to long-term decisions. Chapter 4 offers a first compilation of some of these initiatives in the EU, which can be very inspiring for others, irrespectively of their geographic and sectoral backgrounds.

In conclusion, the contents and findings of this book, as summarized in Chap. 5, are well aligned with the spirit of the EU Strategy on Adaptation to Climate Change, adopted in early 2013. The book aims, as does the CIRCLE-2 project from which it emanates, to assist informed decision-making, and it effectively provides added value through increased knowledge sharing across the EU, contributing with a valuable insight on how to deal with the climate uncertainty challenge.

Brussels   Humberto Delgado Rosa

Preface

Adaptation to climate change has gained substantial political momentum in recent years, both globally and in Europe. Focus was initially on the needs of vulnerable developing countries, but the human and economic impacts of recent extreme climatic events in industrialized countries have emphasized that adaptation is needed everywhere.

Dealing with the large array of uncertainties related to climate and climate change has been acknowledged as a key challenge for adaptation decision-making at all levels; it is a growing area of interest both in the academic literature and in "real-world" practice. Decision-making on adaptation is more and more often required to account for the complexity associated with climate change science, and academics and practitioners alike are demanding clear and coherent guidance on how to recognize, interpret, and communicate uncertainties.

CIRCLE-2[1] has responded to this need by founding the Joint Initiative on Climate Uncertainties.[2] This initiative is a coordinated transnational effort, within the scope of CIRCLE-2, aimed at sharing and advancing scientific knowledge and practice on dealing with and communicating climate and climate change uncertainties in support of adaptation decision-making. It has established a network of renowned excellence, capable of sharing and advancing knowledge and practice on the topic. It also has the stated objective of producing a publication intended to serve as a "guide" to uncertainty in adaptation decision-making that is able to provide practical case study examples where dealing with uncertainties was successfully accounted for (or identified but failed).

A growing body of literature describes new methods and tools, presenting innovative ways of treating uncertainty in decision-making processes, mostly from a theoretical point of view or describing an individual case. However, practical

---

[1] CIRCLE-2 is a European Network of 34 institutions from 23 countries committed to fund research and share knowledge on climate adaptation and the promotion of long-term cooperation among national and regional climate change programs. More information at http://www.circle-era.eu

[2] http://www.circle-era.eu/np4/P_UNCERT.html

guidance describing how the methods and tools have been applied to inform actual decision-making processes and the kind of results they have yielded is equally important. Lessons learned from practical experience can deliver substantial added value, but must be viewed in context. For example, differences in the institutional setup, in the time horizon and reversibility of adaptation decisions, in the predictability of relevant climatic changes, and in the relative importance of climate change compared to other factors are all important issues when taking adaptation decisions.

This book is targeted specifically at policy developers and advisors, practitioners, climate knowledge brokers, researchers, and interested climate change adaptation decision-makers. It differs from other titles addressing climate change adaptation and uncertainty since it uses real-life cases to derive "guidance" or "lessons learned," aimed at helping decision-makers and their advisors to address pertinent uncertainties in actual adaptation situations. To this end, the book includes an overview of adaptation information at the national level in Europe and a compilation of practical case studies and consequential "lessons learned" in Europe, with further examples from Canada and New Zealand.

We hope you find this to be a useful publication and that you enjoy reading it!

Lisbon, Portugal

Tiago Capela Lourenço
Ana Rovisco

Acknowledgements

This book is about sharing experiences and lessons learned on making adaptation decisions that include consideration of uncertainties. It is a result of the work developed by the CIRCLE-2 Joint Initiative 'Climate Uncertainties', between 2010 and 2013.

The work was financially supported by the Fundação Calouste Gulbenkian (Portugal) to whom the editors would like to express their gratitude.

We would also like to thank all our institutions for kindly providing support and the necessary human resources to carry out this initiative.

We would like to particularly acknowledge all those that contributed with case-studies for their time and availability to discuss their work with us.

We wish to acknowledge all those who participated in the CIRCLE-2 workshops that lead to this initiative and in following ones, for sharing their knowledge with us.

We are extremely grateful to all chapter reviewers for their support and for their very important contributions in this specific area of work.

We would like to express our utmost appreciation to the overall book reviewers, Bob Webb, Filipe Duarte Santos and Rob Swart, that read the entire manuscript from cover to back and provided a priceless contribution to our coherency.

Finally, we would like to leave a very special word of appreciation and gratitude to Helen Colyer and Ana Gomes for their tremendous work and support. Your efforts have made this the positive contribution that we foresaw.

To all a very sincere thank you.

Tiago Capela Lourenço
Ana Rovisco
Annemarie Groot
Carin Nilsson
Hans-Martin Füssel
Leendert van Bree
Roger B. Street

Editors

Tiago Capela Lourenço | Ana Rovisco | Annemarie Groot | Carin Nilsson
Hans-Martin Füssel | Leendert van Bree | Roger B. Street

Tiago Capela Lourenço
Faculty of Sciences, CCIAM (Centre for Climate Change Impacts, Adaptation and Modelling), University of Lisbon, Ed. C8, Sala 8.5.14, 1749-016 Lisbon, Portugal
e-mail: tcapela@fc.ul.pt

Ana Rovisco
Faculty of Sciences, CCIAM (Centre for Climate Change Impacts, Adaptation and Modelling), University of Lisbon, Ed. C8, Sala 8.5.14, 1749-016 Lisbon, Portugal
e-mail: acrovisco@fc.ul.pt

Annemarie Groot
Alterra – Climate Change and Adaptive Land and Water Management, Wageningen University and Research Centre, Droevendaalsesteeg 3A, 6708 PB Wageningen, Gelderland, The Netherlands
e-mail: annemarie.groot@wur.nl

Carin Nilsson
Centre for Environmental and Climate Research, Lund University, Sölvegatan 37, S-223 62 Lund, Sweden
e-mail: carin.nilsson@cec.lu.se

Hans-Martin Füssel
Air and Climate Change Programme, European Environment Agency, Kongens Nytorv 6, 1050 Copenhagen K, Denmark
e-mail: martin.fuessel@eea.europa.eu

Leendert van Bree
Department of Spatial Planning and Quality of Living, PBL Netherlands Environmental Assessment Agency, Oranjebuitensingel 6, 2511 VE The Hague, The Netherlands.
e-mail: leendert.vanbree@pbl.nl

Roger B. Street
UKCIP, Environmental Change Institute, University of Oxford, South Parks Road, Oxford OX1 3QY, UK
e-mail: roger.street@ukcip.org.uk

Other contributors

Mikael Hildén Climate Change Programme, SYKE, Finnish Environment Institute, Mechelininkatu 34a, PO Box 140, FI-00251 Helsinki, Finland

Jeroen van der Sluijs Department of Innovation, Environmental and Energy Sciences; Faculty of Geosciences, Utrecht University – Copernicus Institute of Sustainable Development, Heidelberglaan 2, 3584 CS Utrecht, The Netherlands

Reviewers

Filipe Duarte Santos | Rob Swart | Robert Webb

Filipe Duarte Santos
Professor of Physics and Environmental Sciences at the University of Lisbon, Director of the Research Center SIM – Systems, Instrumentation and Modeling for Space and the Environment, University of Lisbon, Faculty of Sciences, CCIAM (Centre for Climate Change, Impacts, Adaptation and Modelling), Ed. C1, Sala 1.4.21, 1749–016 Lisbon, Portugal

Rob Swart
Coordinator of international climate change adaptation research, Earth Systems Science Group – Climate Change and Adaptive Land and Water Management, Wageningen University and Research Centre, Droevendaalsesteeg 3A, Building 100, 6708 PB Wageningen, The Netherlands

Robert Webb
Senior Fellow, Fenner School of Environment and Society, Program Leader, Leading Adaptation Practices, Australian National University, Canberra ACT 0200 Australia

Copy-editor

Helen Colyer
e-mail: hcolyer@googlemail.com

List of Case-Study Authors

4.2.1. Water Supply Management in Portugal

Authors: David Avelar (University of Lisbon, Faculty of Sciences, Centre for Climate Change, Impacts, Adaptation and Modelling), Tiago Capela Lourenço (University of Lisbon, Faculty of Sciences, Centre for Climate Change, Impacts, Adaptation and Modelling) and Ana Luis (EPAL – Empresa Portuguesa das Águas Livres, SA)
Contact details: dnavelar@fc.ul.pt

4.2.2. UK Climate Change Risk Assessment

Author: Helen Udale-Clarke (HR Wallingford)
Contact details: h.udale-clarke@hrwallingford.com

4.2.3. Water Resources Management in England and Wales

Authors: Ana Lopez (Centre for Climate Change Economics and Policy, Grantham Research Institute, London School of Economics) and Glenn Watts (UK Environment Agency)
Contact details: ana.lopez@univ.ox.ac.uk; a.lopez@lse.ac.uk

4.2.4. Water Supply in Hungary

Authors: Agnes Tahy (National Institute for Environment) and Zoltan Simonffy (Budapest University of Technology and Economics - Departament of Sanitary and Environmental Engineering)
Contact details: agnes.tahy@neki.gov.hu and simonffy@vkkt.bme.hu

4.2.5. Climate Change and Health in The Netherlands

Authors: Arjan Wardekker (Health Council of the Netherlands) and Jeroen van der Sluijs (Utrecht University)
Contact details: arjan.wardekker@gmail.com

4.2.6. Flood Risk in Ireland

Author: Conor Murphy (National University of Ireland, Maynooth)
Contact details: conor.murphy@nuim.ie

4.2.7. Coastal Flooding and Erosion in South West France

Author: Bertrand Reysset (National observatory on the effects of climate change - ONERC, France)
Contact details: bertrand.reysset@developpement-durable.gouv.fr

4.2.8. Québec Hydro-Electric Power

Author: Marco Braun (Ouranos), René Roy (Hydro-Québec-IREQ) and Diane Chaumont (Ouranos)
Contact details: braun.marco@ouranos.ca

4.2.9. Austrian Federal Railways

Author: Andrea Prutsch (Environment Agency Austria)
Contact details: andrea.prutsch@umweltbundesamt.at

4.2.10. Dresden Public Transport

Authors: Julian Meyr and Edeltraud Guenther (Technical University of Dresden)
Contact details: ema@mailbox.tu-dresden.de

4.2.11. Hutt River Flood Management

Author: Judy Lawrence (New Zealand Climate Change Research Institute, Victoria University of Wellington)
Contact details: judy.lawrence@vuw.ac.nz

4.2.12. Communication of Large Numbers of Climate Scenarios in Dutch Climate Adaptation Workshops

Author: Luuk Masselink (Wageningen University)
Contact details: luuk.masselink@wur.nl

Contents

1. **Introduction to the Use of Uncertainties to Inform Adaptation Decisions** 1
   Roger B. Street and Carin Nilsson

2. **Background on Uncertainty Assessment Supporting Climate Adaptation Decision-Making** 17
   Leendert van Bree and Jeroen van der Sluijs

3. **How Is Uncertainty Addressed in the Knowledge Base for National Adaptation Planning?** 41
   Hans-Martin Füssel and Mikael Hildén

4. **Showcasing Practitioners' Experiences** 67
   Annemarie Groot, Ana Rovisco, and Tiago Capela Lourenço

5. **Making Adaptation Decisions Under Uncertainty: Lessons from Theory and Practice** 139
   Tiago Capela Lourenço, Ana Rovisco, and Annemarie Groot

Key Terms 163

References 171

Book Reviewers 173

Index 179

Chapter 1
Introduction to the Use of Uncertainties to Inform Adaptation Decisions

Roger B. Street and Carin Nilsson

Decisions and policies that address existing and future risks and opportunities are necessary and constantly taken. We depend on these. Our social, cultural and economic sustainability and that of future generations are determined by the quality of these decisions and the appropriate use of the evidence that informs them. Uncertainty is associated with limitations on the knowledge that is the basis of that evidence and it is intrinsic to science where questions typically arise as to what information can be considered valid and reliable. As uncertainties are inherent in such evidence, and are in many cases irreducible, they must be included in decision-making processes.

In evidence-based adaptation decision support, uncertainty can be associated with the choice of socio-economic scenarios, climate models, biophysical impact models, integrated assessment models, vulnerability assessments, and appraisal of adaptation options and policies.

For example, how much will the sea level rise in the future 100 years, and how many people and what infrastructures will be located near the coast during this period? How can we plan and design when the projected sea level rise and population from different sources of information provide different estimates and each include different assumptions and uncertainties? The differences in evidence, including the associated uncertainties, do not need to be seen as a barrier to action.

R.B. Street (✉)
UKCIP, Environmental Change Institute, University of Oxford,
South Parks Road, Oxford OX1 3QY, UK
e-mail: roger.street@ukcip.org.uk

C. Nilsson
Centre for Environmental and Climate Research, Lund University,
Sölvegatan 37, S-223 62 Lund, Sweden
e-mail: carin.nilsson@cec.lu.se

They do not diminish the need for action nor require a delay in action to some future time or generation. In summary, they are not an insurmountable obstacle to decision-making.

Experience suggests that it is better to acknowledge and embrace these uncertainties. They need to be managed and effectively incorporated in decisions and policies, there making them more robust as they are based on the available evidence, including the uncertainties. Experience also recognises that, in the context of adaptation to climate change, ignoring the uncertainties in the evidence or limiting consideration to the bias of a known or desired comfort zone increases the risk of maladaptation with potentially high social, economic and environmental costs.

It is our intention to share the experiences and knowledge of others to enable the reader to address these challenges. The book provides insights and background information to inform decisions and policy-making processes, with a special attention on how to include information on a changing climate in planning and implementing the adaptation needed by society to meet the challenges ahead.

We hope that it will be useful and inspire further learning and sharing of experiences.

1.1 Why Is Guidance on the Role of Uncertainty Needed and Who Is it for?

"Guidance is needed, to be able to choose from the scatters of data which is around. You need help if you are not a researcher. There is a need for a guidance which is practical. It should not do the choices for us, but support us to do the right choice."

(Engineer Bengt Rydell, from the Swedish Geotechnical Institute, working with risk investigations for landslides and slope instability at the local and regional level in Sweden together with climate consultants on adaptation planning)

"Clear guidance is essential in helping decision-makers understand what is meant by uncertainty in their own specific context."

(Peter Walton, Oxford University)

1.1.1 Purpose of the Guidance

In presenting this publication we have decided to emphasise the sharing of experience, knowledge and lessons learned. This is done primarily through the presentation of experiences and lessons learned by those taking decisions and developing policy.

This intends to provide support for those navigating their way, or considering doing so, through using the myriad of data and information and in communicating their decisions. The choices made (such as which data to use, how to use it in the process, or how to collect new data) and their impacts on decisions and policies, should involve the consideration of the associated uncertainties.

We have purposely tried to avoid being prescriptive with a preference for being informative.

By choosing case studies to inspire and inform we sought to provide:

- Evidence that there are those in different sectors and countries who have experience in managing uncertainty and that their lessons learned are of value and informative beyond their specific project;
- Examples that would demonstrate how others have used engagement between decision-makers and providers of the required evidence.

We hope that the reader will want to explore these experiences and lessons out of curiosity and a desire to learn. We aim to provide the reader with insights into the following questions:

- What have others done in my sector or my country?
- How did they manage and communicate the uncertainties?
- What are the assumptions and reasoning behind the approach taken?
- Can I adopt a similar approach in my situation?

1.1.2 Who Should Be Using the Guidance?

The information available, including the case studies, has been developed to inform decision-makers as well as policy-makers, advisors and practitioners at the International, European, national, regional and local levels within the private and public sectors.

We recognise that the intended audience for this publication is not a homogenous group, but rather a broad spectrum in terms of capabilities and interest. The probable common denominator is that they are 'dealing with' evidence that includes uncertainties and with its consequences in making and communicating decisions.

The audience might include a practitioner who works in national or local government and need to rethink planning issues in relation to flood or other risks, or someone at the regional level looking for ways to support their regional adaptation plan. Some other examples of those that might find the information provided useful are a business sector manager tasked to consider climate change in the context of business development or continuity or a policy developer/analyst at the national level who needs to understand alternative ways to deal with information to support adaptation decisions.

The book targets also a scientist who would like to know more about the role of evidence in decision-making, or an engineer implementing an adaptation action.

Even though this book is not specifically intended for the public, we hope it will be an inspirational read for those informing them via the media and other means of communication.

1.1.3 Why Is the Guidance Needed?

Including uncertainty along with evidence, in the form of data and information, can be complex and challenging. Both users and providers of data and information have expressed frustration when trying to understand and communicate the different means of incorporating uncertainties.

Adaptation decision-making requires information on risks and vulnerabilities in order to identify needs and adaptation options that are able to build capacity and reduce risks. But is it necessary to have certain information, or 'an accurate estimate', to successfully plan for adaptation?

Within CIRCLE-2,[1] participants in several workshops[2] have expressed an urgent need for concise, practical guidance to address these frustrations. They have asked for advice on managing and characterising uncertainties in the evidence, including on how the uncertainties and their inclusion as part of the evidence relates to the specific framing of the decision and the broader utilisation of the evidence.

This understanding involves exploring uncertainties with the intention of framing them to support and inform decisions and begins with understanding the impact of uncertainties on the decision-making process and the need to retain credibility and legitimacy of the process and resulting decisions.

Guidance is needed as there is not a single, one-size-fits-all method for managing uncertainty. The methods used should reflect the specific situation, the evidence considered; the decision framing and characteristics of how and why the evidence was used (e.g., risk tolerance).

There is also a continuing need to reconsider how we describe, communicate and use evidence, including associated uncertainties. Our understanding of how we use evidence to inform should evolve alongside changes in the science and use of that evidence.

> "I would like a guidance that is able to update my knowledge related to including uncertainties in my analyses and that is able to keep me up-to-date on this topic"
>
> (Anna Bratt, PhD in Environmental Science and the regional coordinator at the County Administrative board of Östergötland, Sweden, with the task to coordinate adaptation within the county Östergötland)
>
> "I would expect the guidance to written in such a way that it can be used throughout the decision-making process, the beginning, middle and for any review/evaluation process".
>
> (Peter Walton, Oxford University)

---

[1] CIRCLE-2 is a European Network of 34 institutions from 23 countries committed to fund research and share knowledge on climate adaptation and the promotion of long-term cooperation among national and regional climate change programmes. More information is available at http://www.circle-era.eu

[2] Workshop 'Dealing with Uncertainties in Climate Change Impacts, Vulnerabilities and Adaptation Research', Nov 2010, Stockholm, Sweden; Workshop 'From National Adaptation Strategies to Concrete Adaptation Actions – Good Practice Examples', Oct 2011, Vienna, Austria; Workshop 'Supporting the Development of the EU Strategy for Adaptation to Climate Change – Views and Challenges in Eastern Europe', June 2012, Vienna, Austria.

1.2 Why Is it Important to Include Uncertainties in Adaptation Planning?

The very simple answer to *why* it is important to take uncertainty into account in adaptation is that it strengthens decisions and their relevance. But what is really meant by this? How does it work?

One approach to answering these questions is to consider what it would mean if we did not account for uncertainties. Effectively we would not be considering all the evidence and would risk incurring unexpected consequences from our decisions. Without considering the full range of possibilities we would risk maladaptation, including over or under adaptation as the range of possible future had not been considered. There is an increased risk of being unprepared and caught unaware. At best the consequences may be purely financial, but there could also be a loss of property and livelihood, social and economic insecurity and inequity, loss of environmental services and even loss of life.

Uncertainty can be an aid to informed decision making, necessary on the path to successful and sustainable adaptation. This assumes, however, that these uncertainties are known and their effect on decisions and therefore how they should be considered are understood.

What must be known about uncertainties in order to incorporate them into decisions? Simply using just a single value or some limited set of data – for example 'an increase in peak river flow by 10 %' indicative of the existing or future state at some location – although potentially easy to use, may not provide sufficient information about the true or possible future state(s).

If the goal is to make a decision on future action (policy or practice), knowledge of the possible future in which those actions will be operating is essential. This should include evidence on the possible future state(s), including the assumptions and limitations behind that evidence that will inform how that evidence can and should be used.

As such, when using evidence it is important to understand *the nature of the knowledge related to that evidence*. What is known about the evidence and what is unknown or uncertain?

In Chap. 2 the location, level, and nature of uncertainties are further explored as well as ways to use uncertainty assessments to guide the decision-making process on climate adaptation.

1.2.1 Why Cannot Decisions Wait Until Uncertainties are Resolved?

Deciding not to act, based on a desire to wait until uncertainties are reduced or based on fear of making a decision when there are uncertainties, may not be viable or acceptable. Research cannot reduce all uncertainties and in some cases can even increase them. Despite existing efforts to reduce uncertainties, prospects of eliminating them are limited.

There are those that are essentially irreducible as they are associated with the chaotic nature of systems and their interactions. They are unpredictable or occur as a result of change.

In addition, new uncertainties can arise as a result of a better understanding of the system of interest (new understanding reveals aspects or characteristics that were previously unknown). This means that decision-making processes will always be required to deal with the uncertainties present.

> "Uncertainty has become a pejorative term that has begun to be used as a reason for inactivity / delaying decision-making rather than accepting it as part of a normal decision-making process."
>
> (Peter Walton, Oxford University)

1.2.2 Why Is Considering Uncertainty Important?

Users have expressed a fear of making the wrong decisions or reluctant to be open-minded about how to use the available evidence. There is a tendency for decision making to justify the retention of the status quo and old habits and the use of uncertainty as a reason for inaction. The emphasis on uncertainty within the scientific community often enhances that reluctance or fear, rather than empowering decision-makers to use the available evidence to their advantage.

Following is a summary of some reasons why it is important to consider uncertainty in decision-making:

- **Uncertainty is inherent.** Consideration of uncertainty is an essential element of decision-making as it is inherent in all evidence. It is an integral part of supportive data and information, especially but not only in that related to the future. Appropriately integrating the associated uncertainties as part of the evidence provides a better understanding of that evidence and can enhance its utility within decision-making processes.
- **More relevant and robust decision-making.** Recognising the nature and characteristics of uncertainty and reflecting these in how the associated evidence is used are crucial to making more relevant and robust decisions. By acknowledging and considering uncertainties, rather than expecting readily identifiable and deterministic outcomes, the uncertainties become more manageable. As a result, it becomes possible to formulate coherent decisions and policies.
- **Minimise the potential for maladaptation.** Not 'sufficiently' including uncertainties increases the likelihood that the action taken will be inadequate, inappropriate or increase vulnerability. There is an increased likelihood of maladaptation when using information which does not incorporate uncertainties.
- **Ignoring uncertainty conceals risks.** Ignoring uncertainty can undermine effective risk management as the risks that would result from including uncertainty are simply ignored and not considered in actions to be taken. Uncertainty about climate change science and policy options is often used as an excuse for inaction or is ignored to simplify policy debates.

Some would suggest that it is easier not to incorporate uncertainties and address only that which is more certain and let future generations deal with the results. However, there is a need to recognise the consequences of focusing on one or a limited range of scenarios, while a large range would be required to capture future possibilities.

There is also a common human tendency to dismiss uncertain consequences as not urgent, even if the consequences are potentially severe. There may be a desire to allow time for science to reduce uncertainties before investing financial resources into solutions that may prove unnecessary and investing time into complicated policy and political debates.

Deciding not to act or taking a 'wait-and-see' approach may be an appropriate decision, but that decision should be evidence-based. This means including an evaluation of the risk of not acting or delaying (e.g., considering of the acceptability of any residual risks and the social and economic costs and benefits now and in the future) relative to risk tolerance.

It needs to be acknowledged that decisions under uncertainty always include subjective evaluations of the available knowledge base.

1.2.3 How Can Uncertainty Be Managed?

Throughout the book you will find examples of how others have managed the uncertainties they encountered to enhance the quality of their decisions.

This experience and the sharing of lessons learnt are a key feature of this book. Drawing on these suggests that when integrating uncertainties within a decision-making process there should be a focus on the uncertainties that really matter i.e. those that are relevant to the decision. Are there particular sensitivities or thresholds? Fretting over details and uncertainties that are not relevant to the decision at hand simply enhances the perception of uncertainty and can lead to paralysis. Learning from others can help in this process.

Furthermore, experience has shown that rethinking how uncertain information is used in the decision-making process can be beneficial. Recent reports suggest that there are limits to the usefulness of classic risk analysis for climate-related problems (see Suggested Reading). Hence, seeking robust strategies may prove a preferable approach, and any such analysis, including how information and its associated uncertainties should be embedded in processes that include stakeholder engagement.

This also means considering the framing of the decision- and policy-making process, and consideration of the temporal nature of the uncertainties relative to the temporal aspects of the decision or policy.

Experience also suggests that when communicating the results of a decision-making process there should be a focus on approaches that more effectively characterise and communicate the role of uncertainty. This means that communication should go beyond that used within the scientific community to that required to reach and inform those that are (or should be) engaged in the development and delivery of adaptation.

Based on this experience, effective approaches appear to be those that:

- Explore a wide variety of relevant uncertainties;
- Connect short-term 'targets' to long-term goals over time;
- Identify the risks of failure of proposed options;
- Commit to short-term actions while keeping options open in the mid- to long-term;
- Continuously monitor and evaluate, taking further action when necessary.

1.3 What Information Is Included in This Book and Where Can I Find it?

This publication has been compiled and structured to provide practical examples and background information related to uncertainty and its use in decision- and policy-making. Together, the following chapters are intended to answer those questions being asked.

The table below (Table 1.1) aims to provide a quick overview of the information and examples we have included and where it can be found. The intention is that this table, along with the suggested navigation pathways in Sect. 1.4, will help you make better use of this guidance to address your specific knowledge and evidence needs and better understand how to include uncertainty in your decisions.

1.4 How Can This Publication Be Used?

As there is a diversity of users, there is also a diversity of ways that this publication can be used. We recognise that not all or even many will read the publication from start to finish although we suggest it would be useful. Time availability will often be a limiting factor and many will want to focus on extracting lessons learnt that will meet their specific needs. To this end, a variety of pathways within this publication can be explored to extract relevant lessons and supportive information.

The pathway(s) chosen by each reader will depend on their specific interests (e.g., nature and scope of decisions to be made) and reasons for better understanding the use of evidence that includes uncertainty.

They may be based on a desire to draw on the lessons learnt by others with similar interests, capabilities and challenges; to draw on the lessons learnt and information available to enhance your capabilities to appropriately include and communicate

1 Introduction to the Use of Uncertainties to Inform Adaptation Decisions

Table 1.1 Book overview and respective information of what can be found in each chapter

	Title	Short overview	Useful for questions such as
Chapter 1	**Introduction to the use of uncertainties to inform adaptation decisions**	Purpose and scope. Structure of publication and navigation.	Why should I include uncertainty in my decisions? What will I find in this publication? Why and how should I use this publication? Where will I find information and practical examples that are of interest to me?
Chapter 2	**Background on uncertainty assessment supporting climate adaptation decision-making**	Climate variability, climate change and relationship with development. Uncertainties in climate change and uncertainty typology. Methods of assessing uncertainty and decision-making frameworks. Using uncertainty assessment in adaptation practice. Communicating uncertainty assessment to policy-makers and decision-makers.	Why should I be concerned about climate change in terms of decisions and policy? How does climate variability affect my decisions and how does that differ from consideration of climate change? How do I identify and characterize the uncertainties that I should be considering? What approaches for adaptation planning would I be using and what does this mean for how uncertainty is included? What uncertainties do I need to include? How can uncertainty inform the selection and assessment of adaptation options? How do I communicate decisions and policies that include uncertainties?
Chapter 3	**How is uncertainty addressed in the knowledge base for national adaptation planning?**	Overview of how uncertainties are addressed within nationally used climate change projections, non-climatic scenarios and climate impact, vulnerability and risk assessments in Europe.	Which information and support is available at the national level to assist adaptation planning? Which sources of uncertainty are considered in nationally used climate change projections, non-climatic scenarios and climate impact, vulnerability and risk assessments?

(continued)

Table 1.1 (continued)

	Title	Short overview	Useful for questions such as
		Overview of how these uncertainties are communicated in pertinent reports and web portals.	Which scenarios and models have been used to provide relevant information?
		Guidance for adaptation planning under uncertainty.	Is information on projected changes in climate and socio-economic factors freely and easily available to adaptation planners?
			How is uncertainty about future climate change communicated to adaptation planners, and how are they supported in interpreting and using this information?
			Are there legal obligations to perform climate risk assessments, and to consider uncertainties therein?
Chapter 4	**Showcasing Practitioners' Experiences**	Description of 12 real-life cases showing how policy-makers, decision-makers and researchers have addressed the challenges of using information that includes uncertainties in the context of adaptation.	
		Highlights the challenges faced the types of uncertainties addressed, methods that were used to deal with uncertainties and some of the decision made.	
		Demonstrate how the process of conscientiously addressing climate uncertainties has affected decisions.	
	Case	**Sector key words and appetizing points**	
	Water Supply Management in Portugal (4.2.1)	Water (supply) management and infrastructure:	
		How to build trust necessary to deal with different views and to understand and access decision-making processes.	
		Transferability and sharing of knowledge and data were critical.	
		Quantifying cumulative uncertainty works if properly agreed and communicated from the start.	

UK Climate Change Risk Assessment (4.2.2)	Across several sectors: How to identify robust adaptation options. How to communicate uncertainties. How to build flexibility in planning that allowed for the introduction of new knowledge and information.
Water Resources Management in England and Wales (4.2.3)	Water management: What are the implications of using a limited storyline of future changes How to address the need for water management options that are flexible and robust under a range of possible futures. Avoiding poor adaptation decisions when using climate information within a decision-making framework.
Water Supply in Hungary (4.2.4)	Water management and infrastructure: How one can make decisions despite uncertainty in long-term trends of precipitation. Involving local stakeholders in formulating future demand in drinking water.
Climate Change and Health in The Netherlands (4.2.5)	Water management and health: How to use a questionnaire to gather information on what hinders a more precise analysis of the uncertainty. How the use of uncertainty typology can lead to improved policy decisions.
Flood Risk in Ireland (4.2.6)	Infrastructure and disaster risk reduction: How to consider the performance of adaptation options in the context of uncertainties. How to communicate uncertainties such that decision can be more robust.
Coastal Flooding and Erosion in South West France (4.2.7)	Biodiversity and nature, coastal management, infrastructure and disaster risk reduction: The functions of low/no regret options. Making meaningful coastal investments in the absence of precise predictions. The importance of community involvement.
Québec Hydro-Electric Power (4.2.8)	Water management and infrastructure: How addressing uncertainty can change how scenario information is used. The importance of communications and knowledge exchange in the decision-making process.

(continued)

Table 1.1 (continued)

	Title	Short overview	Useful for questions such as
	Austrian Federal Railways (4.2.9)	Infrastructure: How trend analysis can help in managing uncertainties. The importance of informed engagement and clear communication in the decision making process.	
	Dresden Public Transport (4.2.10)	Infrastructure: Identifying influencing factors (climate and non-climate) and their dependencies across the business as a basis for informing the decision-making process and targeting actions. Stakeholder engagement in the assessment of risks and adaptation options.	
	Hutt River Flood Management (4.2.11)	Infrastructure and disaster risk reduction: How understanding the limits of static flood protection and emergency planning in the context of an uncertain future can inform robust decision making. The role of visual depictions in communicating the effects of climate change uncertainties.	
	Communication of Large Numbers of Climate Scenarios in Dutch Climate Adaptation Workshop (4.2.12)	Water management, agriculture and forestry, biodiversity and nature, and infrastructure: How the method of presentation of climate change information can affect understanding by those making decisions. The role of visualisation in enhancing the understanding of uncertainties and decisions needed to be made.	
Chapter 5	**Making adaptation decisions under uncertainty – lessons from theory and practice**	Presentation of a Common Frame of Reference for analysing and dealing with uncertainties. Analysis of the cases – overview of methods used and the effect on the decision Key findings and guidance on working with uncertainty in adaptation decision	How to analyse uncertainties across the different cases and how do these relate to my own situation? What types of decision objectives were aimed at by the cases? What methods were used to deal with uncertainty in each case? What types of uncertainties did the cases dealt with? What is the summary of the key findings in the cases? What guidance and valuable lessons can I extract from these experiences?
Key Terms	**Acronyms and key terms**	Explanation of acronyms and key terms used within the publication.	What is meant by....?

decisions that include uncertainties; or to realise learning objectives related to working with uncertainties. The following pathways (Figs. 1.1, 1.2, 1.3 and 1.4) are illustrative:

Pathway 1 – Learning from the case studies

Readers looking to take specific adaptation decisions could begin with a specific case study (Chap. 4) in a sector of interest and/or that addresses a similar problem. Terms can be clarified in Key Terms. Together these can enhance understanding of the case study and the applicability of the lessons learnt. The path could end there, but may go on to explore lessons from other case studies.

Pathway 2 – Seeking clarity on the terms used

The reader's path begins by exploring specific terms in Key Terms, then to Chap. 2 to better under the concepts and background information. The reader could then continue by making reference to a specific case study (Chap. 4) to enhance understanding of the concepts and of the specific question that prompted this pathway.

Pathway 3 – Uncertainty in adaptation strategies

The reader's path begins with Chap. 3 with exploring national adaptation strategies and how uncertainty has been addressed in those strategies. The reader could then either continue to Chap. 2 to clarify concepts and approaches that have been used to assess and communicate uncertainty and then to the Key terms to understand the terms that have been used, or go directly to the Key terms.

Pathway 4 – Guidance and a general overview

The reader's path could begin with lessons learnt and a synthesis of key messages from the practical cases (Chap. 5). The pathway then may lead to an exploration of some of the case studies (Chap. 4) for examples from relevant sectors, to an exploration of concepts and background information in Chap. 2, national adaptation strategies in Chap. 3 and finally Key Terms in end.

Whichever pathway is taken there are opportunities provided to learn from others that have already journeyed and navigated the challenges associated with using evidence that includes uncertainties. Like them, the reader will see that uncertainty need not be a barrier to action when it is understood, appropriately included and communicated.

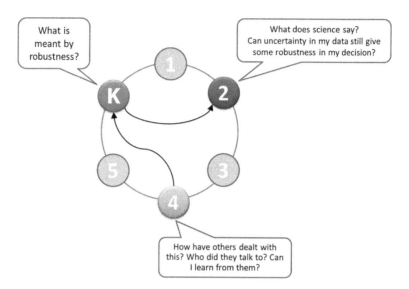

Fig. 1.1 Suggested pathway 1 – learning from the case studies

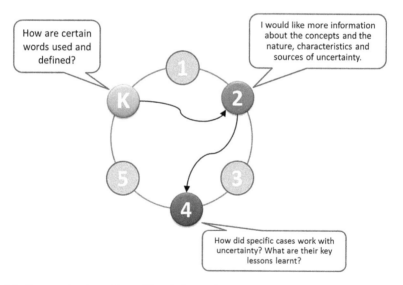

Fig. 1.2 Suggested pathway 2 – seeking clarity on the terms used

1 Introduction to the Use of Uncertainties to Inform Adaptation Decisions

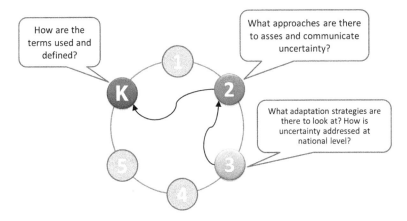

Fig. 1.3 Suggested pathway 3 – uncertainty in adaptation strategies

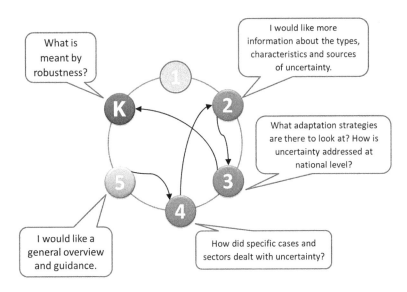

Fig. 1.4 Suggested pathway 4 – guidance and a general overview

Suggested Reading

Dessai, Suraje, and Hulme Mike. 2004. Does climate adaptation policy need probabilities? *Climate Policy* 4(2): 107–128. doi:10.3763/cpol.2004.0411.
CCSP. 2009. Best practice approaches for characterizing, communicating, and incorporating scientific uncertainty in decisionmaking. *In A report by the climate change science program and the subcommittee on global change research*, ed. M. Granger Morgan, Dowlatabadi Hadi, Henrion Max, Keith David, Lempert Robert, McBride Sandra, Small Mitchell, and Wilbanks Thomas, 96. Washington, DC: National Oceanic and Atmospheric Administration.
Dobes, L. 2012. Adaptation to climate change: Formulating policy under uncertainty, CCEP Working Paper 1201, Centre for Climate Economics & Policy, Crawford School of Economics and Government, The Australian National University, Canberra.
Hawkins, Ed., and Sutton Rowan. 2010. The potential to narrow uncertainty in projections of regional precipitation change. *Climate Dynamics* 37(1–2): 407–418. doi:10.1007/s00382-010-0810-6.
IPCC. 2010. Guidance note for lead authors of the IPCC fifth assessment report on consistent treatment of uncertainties. IPCC Cross-Working Group meeting on consistent treatment of uncertainties, Jasper Ridge, 6–7 July 2010.
Lempert, Robert and Nidhi Kalra. 2011. Managing climate risks in developing countries with robust decision making. World Resources Report, Washington DC. Available online at http://www.worldresourcesreport.org.
Lemos, Maria Carmen, and Richard B. Rood. 2010. Climate projections and their impact on policy and practice. *Wiley Interdisciplinary Reviews: Climate Change* 1(5): 670–682. doi:10.1002/wcc.71.
Manning, Martin. 2003. The difficulty of communicating uncertainty: An editorial comment. *Climatic Change* 61: 9–16.
van der Sluijs, Jeroen P., Arthur C. Petersen, Peter H.M. Janssen, James S. Risbey, and Jerome R. Ravetz. 2008. Exploring the quality of evidence for complex and contested policy decisions. *Environmental Research Letters* 3((2): 024008. doi:10.1088/1748-9326/3/2/024008.
van Pelt, S., D. Avelar, T. Capela Lourenço, M. Desmond, M. Leitner, C. Nilsson and R. Swart. 2010. Communicate uncertainties – design climate adaptation measures to be flexible and robust. Proceedings of the CIRCLE-2 workshop on Uncertainties on Climate Change Impacts, Vulnerability and Adaptation, Stockholm, 11–12 November 2010.

Online Resources

On climate-adapt there is a special section called uncertainty guidance, with several links to different ways of dealing with uncertainty: http://climate-adapt.eea.europa.eu/web/guest/uncertainty-guidance-ai.

Chapter 2
Background on Uncertainty Assessment Supporting Climate Adaptation Decision-Making

Leendert van Bree and Jeroen van der Sluijs

Key Messages

- Analysing, characterising, and dealing with uncertainty forms an integral part of establishing and implementing climate adaptation policy.
- The classical elements used in uncertainty assessment (statistics, scenarios and recognised ignorance) can be expanded toward five principal uncertainty dimensions that are crucial for informing/supporting adaptation decision-making: location, level, nature, qualification of knowledge base, and value-ladenness.
- In practice, to deal with uncertainties, but also because of time and budget constraints, uncertainty assessments may follow a three step approach: (1) identify and characterise sources of uncertainty; (2) weigh, appraise, and prioritise uncertainties; and (3) select and apply methods for dealing with uncertainties in decision-making and policy.
- Based on political and societal preferences, adaptation strategies could either use top-down or bottom-up approaches considering adaptation actions based on the best prediction, robustness, or resilience.

(continued)

L. van Bree (✉)
Department of Spatial Planning and Quality of Living, PBL Netherlands Environmental Assessment Agency, Oranjebuitensingel 6, 2511 VE The Hague, The Netherlands
e-mail: leendert.vanbree@pbl.nl

J. van der Sluijs
Department of Innovation, Environmental and Energy Sciences; Faculty of Geosciences, Utrecht University – Copernicus Institute of Sustainable Development, Heidelberglaan 2, 3584 CS Utrecht, The Netherlands
e-mail: j.p.vandersluijs@uu.nl

(continued)

- Adaptation policies that focus on enhancing the system's and society's capability of dealing with possible future changes, uncertainties and surprises (e.g. through resilience, flexibility, and adaptive capacity) seem most appropriate.
- For potential climate-related effects for which rough risk estimates are available, 'robust' measures are recommended.
- For potential climate effects with limited societal and/or political relevance, 'no-regret' measures are recommended.
- For highly relevant potential climate-related effects, precautionary measures can be considered.

2.1 Introduction

Climate affects societies in many ways, and climate variability and climate change are important factors for societal development (Fig. 2.1). Over the past century (1906–2005), global average surface temperatures have increased by 0.74 ± 0.18 °C (IPCC 2007a). Based on observations of global air and ocean temperatures and changes in snow/ice extent and sea level, the Intergovernmental Panel on Climate Change (IPCC) concluded that it is 'unequivocal' that the climate system has warmed (IPCC 2007a).

According to the IPCC, most of the warming since the middle of the twentieth century is *very likely* to be due to the human-induced increase of atmospheric greenhouse gas concentrations. Various climate impacts on both physical and biological systems have been observed. IPCC temperature projections for the end of the twenty-first century range from an increase of 1.1–6.4 °C, compared to end of the twentieth century. These changes in the global average temperature have a wide variety of global, regional and local effects, such as changes in: temperature, sea levels, precipitation and river runoff, drought, wind patterns, food production, ecosystem health, species distributions and phenology, and human health (IPCC 2007b; EEA 2012, 2012a).

At the regional level, changes can, however, substantially differ. For example, the observed Western European increasing temperature trend over the past decades is much larger than the global average trend. Regional climate effects such as changes in atmospheric circulation, and environmental changes such as lower aerosol concentrations, are believed to have played a role in this difference (e.g. PBL 2009). The impacts of expected global changes will differ by region and sometimes by season. In many cases, the impacts will be detrimental, although some regions might welcome some of the changes, provided they remain relatively small; for example, in cold-limited regions warming could be useful for agriculture or access to mineral reserves.

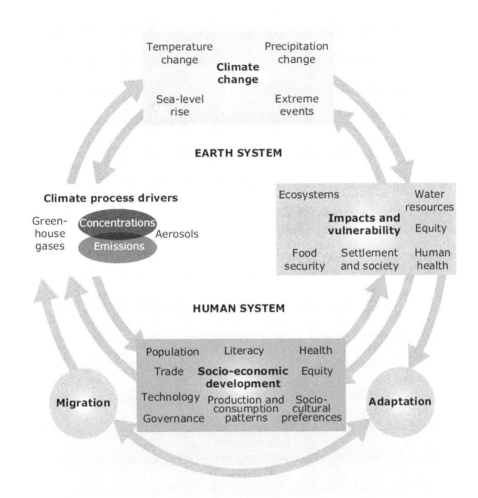

Fig. 2.1 Schematic framework representing anthropogenic climate change drivers, impacts and responses, and their links (EEA 2012a)

Two main responses have emerged in recent decades to deal with climate change: mitigation and adaptation. Mitigation is generally described as "*Limiting climate change by reducing greenhouse gases (GHG) emissions and enhancing sinks*". Adaptation has been described in various ways (Willows and Connell 2003; IPCC 2007b), but they all come down the central issue of "*Adjustments in ecological, social, or economic systems in response to actual or expected climatic stimuli and their effects or impacts. It refers to changes in processes, practices, and structures to moderate potential damages or to benefit from opportunities associated with climate change*" (United Nations Framework Convention on Climate Change: http://unfccc.int/focus/adaptation/items/6999.php).

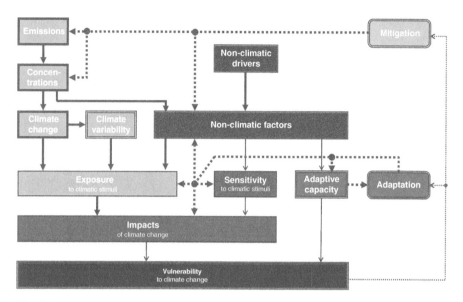

Fig. 2.2 Conceptual framework for climate change impacts, vulnerability, disaster risks and adaptation options (EEA 2012)

Even when taking an optimistic view on the success and timeliness of emission reductions, some degree of climate change is inevitable (e.g. Smith et al. 2000; Dessai and Van der Sluijs 2007; IPCC 2007b), sizeable future emissions will probably remain, and, due to the thermal inertia of the oceans, past emissions have not yet reached their full climate impact.

Adaptation can result in benefits regarding vulnerability to present-day climate and can be economically competitive and attractive. Adaptation measures, however, are seldom taken in response to climate change alone and are often embedded in broader sectorial or integral urban and regional development initiatives (IPCC 2007b; Runhaar et al. 2012). Similarly, in many countries, adaptation strategies address the problem on different spatial scales – that of cities, regions, or on a national scale. They can even be addressed internationally (EEA 2012, 2012a). The adaptation strategies often follow the same format:

- First the reality of climate change is established
- Then there is a scientifically-based analysis of future vulnerabilities and risks on a particular territory (usually based on long-term projections).
- Possible options to counteract these effects are then proposed, and
- Finally these options are assessed in terms of (cost-) effectiveness

This is also reviewed in Chap. 3 for national adaptation strategies.

A widely accepted framework climate adaptation has been developed by EEA (2012) and is presented in Fig. 2.2.

The impacts of climate change are, however, associated with several uncertainties, especially when projections are being made towards the year 2100. These are present in the context of the impact assessment (e.g. in the scenarios and climate data and projections used), and in each step of the assessment itself. They are also cumulative, resulting in an '*uncertainty explosion*' or '*cascade of uncertainty*' (Schneider 1983; Henderson-Sellers 1993; Giorgi 2005; Dessai and Van der Sluijs 2007).

2.1.1 Climate Variability and Climate Change

Whilst the concept of climate change risk is generally acknowledged, there is little apparent distinction made between true (long term) climate change and the short term imperative of responding to climate variability.

The risk is that the "quick-fix", vote-earning, policy responses to climate variability make future adaptation to climate change much harder, less likely, and perhaps even unlikely. For instance, a short-term response to flooding is to provide efficient and effective emergency response and post-disaster support, yet the longer term response should be to reduce the risk through, say, relocation. There has been some policy movement in this direction, for instance managed regression of land on the less populated areas of east coast of the United Kingdom but it has yet to be accomplished within an urban context.

We understand that there is a need for two, yet integrated policy adaptation sets; one for climate variability and one for climate change, which will need different, yet parallel, decision-making processes to be operative. And if possible, there should be clear links between the two. In addition, there is a need for an accurate use of the term "climate change".

To be effective, adaptation should be part of any urban and rural economic development policy and in any related sectorial plans and budgets. We believe that the most important requirements for short-, medium- or long-term decision-making are:

- The policy sets, and
- The projections of climate change risk.

2.1.2 Climate Variability, Climate Change, and Projections of Risks

Climate variability may cause adverse effects like floods, droughts, or intense rainfall/storms. These short-term disruptions could have a significant effect on economies where the economic activity is sensitive to the weather and climate. Policies need to be designed to take sensitivities into account and this is often already the case where they are seamlessly incorporated into a business continuity mind-set of existing governance systems and bureaucracies.

Climate change is on a decadal scale. Very few policies are able to operate on that timescale partly because of the lack of clarity in the objectives and partly because there is a reluctance to commit resources for which there is no political or tangible (near term) return. With these policies the return accrues to a future generation. Individuals are able to plan into the future to a certain extent, by saving a pension for example, but only because they are able to understand the implications of living without a source of income. There is, however, no collective equivalent. An entirely new set of policies must be formulated which have no immediate tangible benefit, being simply a gift to the future.

So, we believe that policies to deal with climate variability and policies to deal with climate change are both needed.

2.1.3 Relationship Between the "Climate" and "Development" Communities

Communities interested in climatic patterns are often distinct from, and do not necessarily "speak the same language" as, those concerned with the economy or resource management. At a minimum, we feel that the understanding between these two constituencies should be improved to establish a common platform for action in areas where the two sets of policy objectives intersect. An example of where progress seems likely is in the factoring in climate change impacts and vulnerabilities when planning for sectoral and overall economic development. Applications range from building institutions for better governance to re-orienting specific investments in physical infrastructure.

How should we enhance climate change adaptation or adaptive capacity through "business as usual" programmes and plans? What are the priorities for investment in adaptation or adaptive capacity, and how should such priorities be determined? These are some of the key questions that need to be answered. Adaptive capacity is the ability to implement adaptations and is a function of such factors as wealth, access to technology, institutional capacity and ability to change.

2.2 Uncertainties in Climate Change

Although trends in climate change are expected to continue, there is considerable uncertainty about the precise rate of change and its concrete impact. Vulnerability to climate change will therefore be greatly affected by the way behavioural, technical, and spatial adaptation strategies and policies are developed and effectively implemented.

A key element in decision-making on climate adaptation is how to deal with uncertainty (Ribeiro et al. 2009). Insight into the uncertainty may determine the preferred adaptation policy in terms of enhancing adaptive capacity, resistance, resilience, robustness or flexibility (Dessai and Van der Sluijs 2007). Models assessing the various sorts of uncertainties to guide policy-makers and decision-makers are therefore crucial instruments for climate proofing (EEA 2012, 2012a).

Decision-making on adaptation under climate uncertainty also involves effective communication and appreciation between science, society, and policy. Such communication and appreciation is often hampered by misunderstandings about the phenomenon of uncertainty in the science and the fundamental limits to climate change and impact predictions.

Lack of systematic attention for unquantifiable uncertainties makes the perceived scientific foundation for climate policies prone to controversies. It can also undermine public support for climate policies, and increase the risk that society is surprised by unanticipated climate changes (Dessai and Van der Sluijs 2007).

The presence of climate uncertainties in adaptation policies challenges all actors in society to assess, evaluate and prioritise adaptation solutions from perspectives such as cost and benefits of investments and short-term and long-term policy preferences. Dealing with complex risks under uncertainty can rarely have a blue-print approach, but does require a tailored and targeted strategy. Because uncertainty assessment is a relatively new scientific discipline, there is significant room for dealing transparently with uncertainty in decision-making and policy. There is also scope for the possible role of other important factors such as ethics (Briggs 2008; Knol et al. 2009).

2.3 Uncertainty Typology

There is a distinction between various sources of uncertainty: decision uncertainty (e.g. related to human decisions that determine future GHG and aerosol particle emissions), natural variability (e.g. related to the internal variability of the climate system), and scientific uncertainty (e.g. related to data gaps, incomplete understanding or insufficient computing power of climate and climate impact models).

An uncertainty typology can be used to classify and report the various dimensions of uncertainty and can improve communication between analysts, policy-makers and stakeholders. It can also help identify where the most (policy) relevant uncertainties can be expected, and how they can be characterised in terms of a number of uncertainty features. Additionally it can serve as a first step of a more elaborate uncertainty assessment, where the extent of uncertainties and their impact on the policy-relevant conclusions are explicitly assessed.

The character of uncertainty is twofold:

- Cognitive – uncertainty in knowledge, and
- Normative – uncertainty in value and goal.

Cognitive uncertainty refers to the level of underpinning and backing of the information (e.g. data, theories, models, methods, argumentation etc.) involved in the assessment of the uncertainty of the problem; it points to the methodological acceptability and the rigour and strength of the employed methods, knowledge and information, and thus it characterises to a certain extent their (un)reliability.

Normative uncertainty relates to the presence of values and biases in the various choices involved e.g. choices concerning the way the scientific questions are framed, data are selected, interpreted and rejected, methodologies and models are devised and used, and explanations and conclusions are formulated etc.

A variety of different types of uncertainty has been defined and used in the literature and in practice. To be pragmatic, in this book we have used an uncertainty characterization originally proposed by Walker et al. (2003) which has been further developed by RIVM/MNP.[1] This three dimension typology, i.e. location, nature and level of uncertainty, can also be expanded to five principal uncertainty dimensions:

- **Location** – the part of the problem in which the uncertainty occurs,
- **Level** – classification on scale from "complete ignorance" to "knowing for certain",
- **Nature** – whether uncertainty is knowledge-based or a direct consequence of inherent variability,
- **Qualification of knowledge base** – evidence and reliability and of information used, and
- **Value-ladenness of choices** – the extent to which choices made in the assessment are subjective.

This classification of uncertainty is quite crucial for a specific uncertain adaptation issue and the choice of transparent and targeted decision-making and policy strategies which try to deal with it. Choices which will be made in the next decades will determine the future level of climate-proofing and the future room for (additional) changes when climate change and its impacts develop at a different rate to that expected. Understanding of these uncertainties will help policy-makers to select appropriate adaptation policies based on societal preferences. We hope this book will help improve climate adaptation decision-making processes and policy-making by analysing, dealing with, and communicating climate uncertainties.

The five uncertainty dimensions are further explained below:

2.3.1 Uncertainty Location

This dimension relates to the part of the problem in which the uncertainty occurs. Five locations can be identified as follows:

- **Context** concerns the scoping and framing of the problem, including deciding what should be inside and outside the system boundaries i.e. delineation of the system and its environment. It also refers to the completeness of the problems involved.

---

[1] This guidance was developed by the Netherlands Environmental Assessment Agency (formerly RIVM/MNP). More information on it guidance can be found at: http://www.nusap.net/downloads/detailedguidance.pdf

- **Data** refers to measurements, monitoring data, and survey data etc. used in the study. It is the category of information which is directly based on empirical research and data gathering. The data used for calibration of the models involved are also included in this category.
- **Model** concerns the model instruments which are employed for the study. This can encompass a broad spectrum of models, ranging from mental and conceptual models to statistical and causal process models etc. which are often implemented as computer models. In principal models are imperfect and do not take into account all the complexities of the system that is being modelled: model structure (relations), model parameters (process parameters, initial and boundary conditions), model inputs (input data, external driving forces), as well as the technical model, which refers to the implementation in hard and software.
- **Expert judgement** refers to contributions to the assessment not covered above, and that have a more typically qualitative, reflective, and interpretative character. As such this input could also be viewed as part of the 'mental model'.
- **Outputs** from a study are the outcomes, indicators, propositions or statements relating to the problem.

The various aforementioned uncertainties on the location axis can be further characterized in terms of four other uncertainty features/dimensions, which are described in the subsequent sections.

2.3.2 Uncertainty Level

This dimension expresses how a specific uncertainty source can be classified on a gradual scale running from 'knowing for certain' to 'no know'. Use is made of three distinct levels:

- **Statistical uncertainties** are those which can adequately be expressed in statistical or probabilistic terms. For example:
  - statistical expressions for measurement inaccuracies,
  - uncertainties due to sampling effects,
  - uncertainties in model-parameter estimates

  This is often the category of uncertainty referred to in the natural sciences. Scientists may implicitly assume that descriptions of the real system being studied are certain, and that the data employed are representative. However, there may be additional forms of uncertainty at play (see below), which can surpass the statistical uncertainty in size and seriousness and which require attention.
- **Scenario uncertainties** are those which cannot be depicted adequately in terms of chances or probabilities, and can only be specified in terms of (a range of) possible outcomes. For these uncertainties it is impossible to specify a degree of probability or belief, since the mechanisms which lead to the outcome are not sufficiently known. Scenario uncertainties are often construed in terms of 'what-if' statements.

- **Surprise/ignorance uncertainties** are those for which existence is acknowledged, but magnitude cannot be established. There may, for example, be limits of predictability and knowledge ('chaos') or unknown processes. This uncertainty level can appear as recognised ignorance ('known unknowns') or total ignorance ('unknown unknowns').

Uncertainties related to a specific location can appear in any of the abovementioned guises: while some aspects can be adequately expressed in 'statistical terms', other aspects can only be expressed in terms of 'what-if' or 'ignorant' statements.

When we consider climate change, the frequencies distributions in climate data from the past cannot be used for guiding the decisions, because they are likely to change. Consequently we need to address scenario uncertainty and ignorance.

2.3.3 Nature of Uncertainty

Is the uncertainty primarily a consequence of the incompleteness and fallibility of knowledge ('*knowledge-related*' or '*epistemic*' uncertainty) or is it due to the intrinsic indeterminate and/or variable character of the system being studied ('*variability-related*' or '*ontic*' uncertainty)? The first form of uncertainty can possibly, though not necessarily, be reduced by more measurements, better models and/or more knowledge; the second form of uncertainty cannot be addressed this way for example, like inherent indeterminacy and/or unpredictability; randomness, or chaotic behaviour of the climate system.

In many situations uncertainty manifests itself as a mix of both forms; there is an unequivocal delineation between 'epistemic' and 'ontic' uncertainty. Moreover a combination of taste, tradition, specific problem features of interest and the current level of knowledge and ignorance with respect to the specific subject determines to a large part where the dividing line is drawn. The choice can however be decisive for the outcomes and interpretations of the uncertainty assessment; It reflects to a large extent the distinction between uncertainties which are 'reducible' and those which are 'not reducible' by means of further research.

2.3.4 Qualification of the Knowledge Base

The qualification of the knowledge base refers to the degree to which the established results and statements are underpinned (i.e. evidence-based). Examples of such results and statements are as follows:

- The policy-advice statement, such as 'the norm will still be exceeded when the proposed policy measures become effective', or 'the total annual emission of substance A is X kiloton'.
- Statements on the uncertainty in the policy statement such as 'the uncertainty in the total annual emission of substance A is …'

The degree of underpinning can be considered as weak, fair or strong. If underpinning is weak, this indicates that the statement of concern is surrounded by much uncertainty, and deserves further attention.

This dimension in fact characterises the qualification of the knowledge base and the reliability of the information (i.e. data, knowledge, methods, arguments etc.) which is used in the assessment. More detail can be found in the tool-catalogue summarised in Sect. 2.4 and van der Sluijs et al. (2003)

2.3.5 Value-Ladenness of Choices

The final dimension for characterising uncertainties describes whether a substantial amount of 'value-ladenness' and subjectiveness is involved in making the various implicit and explicit choices during an assessment. Examples include:

- How the problem is framed *vis à vis* the various views and perspectives on the problem,
- Which knowledge and information (data, models) is selected and applied,
- How the explanations and conclusions are formed and expressed.

If the 'value-ladenness' is high for any part of the assessment, then it is imperative to analyse whether this could lead to an arbitrariness, ambiguity or uncertainty of the policy relevant conclusions. We believe that different views and perspectives in the assessment should then be explicitly dealt with and the scope and robustness of the conclusions discussed in an explicit manner.

2.4 Methods of Assessing Uncertainty

RIVM/MNP have started to develop a tool catalogue,[2] based on the work of Van der Sluijs et al. (2004). This first tool has provided guidance to the character and extent of different sorts of uncertainties in climate adaptation assessments. Later on (Dessai and Van der Sluijs 2007) this catalogue has been further developed into specific techniques that help the user to assess and deal with uncertainties in climate change and adaptation decision-making.

These tools, methods and approaches are listed bellow (no prescribed order) and comprise the list that was applied to the reporting of the real-life cases in Chap. 4:

- Scenario analysis ("surprise-free")
- Expert elicitation
- Sensitivity analysis
- Monte Carlo
- Probabilistic multi model ensemble

---

[2] This tool catalogue can still be downloaded at: http://www.nusap.net/downloads/toolcatalogue.pdf

- Bayesian methods
- Numeral Unit Spread Assessment Pedigree (also known as NUSAP/Pedigree Analysis)
- Fuzzy sets/imprecise probabilities
- Stakeholder involvement
- Quality Assurance/Quality Checklists
- Extended peer review (review by stakeholders)
- Wild cards/surprise scenarios
- For a comprehensive analysis of these methods and their application to adaptation decision-making see Dessai and Van der Sluijs (2007).

Attention should be paid to the fact that both the methods for uncertainty assessment mentioned here and the frameworks for decision-making under uncertainty presented in the next section have different capabilities in the extent to which they can deal with each of the uncertainty typologies described in Sect. 2.3.

In Chaps. 4 and 5 you can find further information on how these methods and frameworks have been applied in practice and how they have contributed to real adaptation decisions.

2.5 Decision-Making Frameworks Under Climate Change Uncertainty

Climate variability is a challenge to the management of risks and uncertainties and may even be amplified by climate change. As such, management depends on the availability of data but it may also be region dependent. Statistical uncertainty can be quantified as a probability density function and can be addressed in policy by a classic risk approach. Some examples are as follows:

- The maximum allowable inundation probability of the urban area in the West of the Netherlands is set to once in 10,000 years. Consequently, the tide with a historical frequency of once in 10,000 years is chosen as the design water-level for determining the level of the dikes and coastal defences.
- The bearing-strength for flat roofs of buildings to be prescribed in the building code can be based on historic data of frequency and amounts of peak snow fall.
- The drainage sewage system in a city can be based on the frequency and intensity of past intense rainfall events to keep the risk of wet feet on an acceptable level.

Future developments of the main drivers of climate change (economic growth and population growth) are inherently uncertain. These can only be explored using projections and scenarios, but the most frequent probability of each scenario is simply unknown. Further, our detailed understanding of the climate system is rather incomplete and all kinds of surprises and unforeseen responses of the climate system and unanticipated impacts may pop up. This is classified as ignorance. The classic risk approach alone is then no longer adequate and needs to be modified

Table 2.1 Different approaches: the spectrum from top-down to bottom-up

Framework	Strategy	Approach
Top-down (predict and quantify changes in stressors)	Act on the best prediction	Based on single scenario
	Robustness-oriented adaptation	Based on range of scenarios Exploratory/discursive
Bottom-up (analyse and reduce vulnerabilities of impacted system)	Resilience-oriented adaptation	Preparing for unknown changes

by approaches that can cope with scenario uncertainty and ignorance. Understanding the relative importance of statistics, scenarios and ignorance in a given adaptation situation is crucial for the choice of a suitable policy strategy to address these uncertainties. This can be different for each particular adaptation problem.

2.5.1 Top-Down and Bottom-Up Approaches

The decision frameworks and analysis tools to deal with uncertainty can be roughly grouped into two schools of thought (see Table 2.1):

- Top-down approach
- Bottom-up approach

The difference between top-down and bottom-up approaches is in the direction in which the causal chain is followed in the reasoning. The top-down approach explores the accumulation of uncertainty from top to down, i.e. from emission scenarios, to carbon cycle response, to global climate response, and to regional climate scenarios. The end result is a range of possible local impacts which enable needs to be anticipated and quantified.

On the other hand, the bottom-up resilience based approach starts at the other end of the causal chain: the impacted system, and explores how resilient or robust this system is to changes and variations in climate variables. It determines how adaptation can make the system less prone to uncertain and largely unpredictable variations and trends in the climate. Resilience also means that the impacted system is suitably adapted to ensure that its essential functions can recover more quickly after a shock. It also ensures quick restoration after damage and rapid response times following early warning signals.

Table 2.1 demonstrates how the different approaches detailed below can be classified on the analysis spectrum. Examples of all types of approach are provided in Table 2.2. For reasons of clarity, the wording 'predict' is also often used as 'project', and the two approaches are used both as providing complementary insights, i.e. not mutually exclusive.

Table 2.2 Top-down and bottom-up climate adaptation examples

	Best prediction	Robustness	Resilience
Flooding	Set flood safety standards based on historical records, or extrapolation of these using a 'best-guess' of the future situation.	Heighten dikes or raise ground level based on national scenarios.	Evacuation and contingency plans.
		Potentially reserve land for further dikes (spatial claims).	Recovery plans. Monitoring and warning systems. Compartmentalisation. Floating (or floatable) buildings. Flood-proof materials for infrastructure and 1st floors of buildings.
Extreme precipitation	Set carrying capacity of flat roofs based on historical records, or extrapolation of these using a 'best-guess' of the future situation.	Set sewer dimension standards to cope with increased and intensified rainfall.	Raised pavements.
	Same for sewer dimensions.		Permeable pavements and/or more soft surfaces (e.g. public or private green). 'Water squares' and similar temporary retention options.
Drought	Design water storage facilities to allow coping with the best estimate for drought occurrence.	Assess the ability of freshwater supply system to cope with range of future circumstances (under current conditions and proposed changes).	Diversify sources for fresh water.
		Change setup and standards for the power supply system to cope with warmer water and lower water tables (for power plant cooling).	Diversify power generation techniques (i.e. include more that do not depend on water cooling).

(continued)

Table 2.2 (continued)

	Best prediction	Robustness	Resilience
Heat waves	Set building standards for isolation, ventilation, and/or cooling options based on expected maximum heat wave in future.	Design cooling systems for buildings to cope with a range of future heat circumstances.	Heat Action Plan with advice and options for staff of senior citizen homes. Increase open water and vegetation in urban areas. Plan orientation of streets/buildings to allow for 'urban ventilation'.

Source: Based on Dessai and van der Sluijs (2007), Wardekker et al. (2010), Runhaar et al. (2012)

Act on The Best Prediction

In some top-down adaptation frameworks, climate change scenarios are considered the main driver of biophysical and socio-economic impacts, thus being of key importance in devising adaptation strategies (Dessai 2005). If policy-makers select a single scenario as the basis for the design of adaptation policies, we call this strategy "*act on the best prediction*". Note that '*best*' does not necessarily refer to '*most likely*' but can also be interpreted as '*considered to be the most relevant for the decision at hand by the policy-maker*'.

Robustness-Oriented Adaptation

Robustness-oriented adaptation strategies focus on climate-proofing to a range of possible futures. That means that the system keeps performing within acceptable limits or can be restored within an acceptable time frame, given the known climate variability, the range of relevant climate scenarios, and considering possible surprises or wild cards. The main strength of these approaches lies in coping with scenario uncertainty.

A top-down way of robustness-oriented adaptation is to use climate scenarios for dimensioning adaptation measures. Internationally, traditional scenario analyses such as those performed by the IPCC (2005) have become an important tool in climate change-related decision-making. At the national and urban scale, some countries and cities have also developed regional climate scenarios. Traditional scenario methods allow for a relatively technocratic approach, using in-house experts or consultants.

Robust decision-making can also include participative approaches with a broader set of stakeholders. Overall, the approach can be used to scope relatively large-scale options and structural measures, as well as for the critical evaluation of proposed options packages.

Resilience-Oriented Adaptation

The other school of thought is resilience-oriented. Resilience is defined as the capacity of a system to tolerate disturbance without collapsing into a qualitatively different, usually undesired state. Some uncertainties associated with climate change are accepted as being irreducible; therefore the emphasis is on learning from past events. This thinking comes from the fields of societal and policy learning, adaptive management for natural resources, and complex adaptive systems research. If uncertainties regarding climate impacts are so big that science is unable to provide any reliable estimates, there might still be enough knowledge to strengthen the general resilience of the impacted system. A resilience approach can make a system less prone to disturbances, and enables quick and flexible responses. Including resilience in climate adaptation will make the adapted system better able to deal with surprises than when using traditional predictive approaches alone.

2.6 Using Uncertainty Assessment in Decision-Making Practice on Climate Adaptation

National and local governments are increasingly seeking building blocks for a resilient climate risk reduction policy. Such as policy needs to be based on more insight into the uncertainty of and vulnerability to climate change in the short and longer term. Adaptation measures are also being increasingly examined in relation to coupling and synergy with various policy areas, such as those of nature, agriculture, urban development, transport and the quality of life. The reduction of greenhouse gas emissions is a component of measures to be considered in relation to climate mitigation policy. For the ultimate policy choices, it is important to acquire a clear picture of the advantages and disadvantages of various packages of adaptation measures, and possible positive or negative feedbacks between various policy fields when uncertainties are taken into account.

Climate change is a relatively slow process. There are long-term impacts on societal restructuring and capital investments are relatively irreversible. Since (some) choices have to be made now, to ensure future climate resilience, flexible policy decisions are required. To develop these, the following factors are necessary:

- Targeted framework,
- Adequate impact and adaptation models,
- Relevant decision-making criteria and adaptation principles, including an uncertainty assessment, and
- Support from all relevant stakeholders.

Decision-making, policy and practice make increasingly use of a structured risk management framework. The usually includes a step-by-step process to help to assess what adaptation measures are most appropriate given the risk management goals and targets. A well-known risk management framework in climate adaptation is the one developed by UKCIP (http://www.ukcip.org.uk/risk). In the steps of identification and appraisal of adaptation options and adaptation strategies, uncertainty assessment, and how to deal with it in adaptation policy, is a crucial process.

Principles for weighing and appraising climate adaptation options and adaptation policies can be condensed into the following five elements:

- ***Risk reduction*** – impact and costs of adaptation options to reduce climate risks, economic and environmental damage, and societal encroachment.
- ***Dealing with uncertainty*** – assessment of uncertainty typology; addressing uncertainty in decision-making frameworks, weighing and appraisal criteria, prioritisation principles, and dealing with uncertainty strategies.
- ***Governance feasibility*** – institutional ability; roles and responsibilities of policy and decision-makers and stakeholders.
- ***Realisation and mainstreaming*** – stakeholder support, equity principle, urgency aspects, implementation time, relevant spatial scale, financial (business) model, 'no-regret' or 'low-regret' adaptation options, and co-benefits of mainstreaming adaptation with other policies.
- ***Monitoring, evaluation, and communication*** – framework, indicator set, and action plan to monitor and evaluate the progress and efficacy of climate adaptation policy.

In the preceding paragraphs we have outlined ways to deal with various types of uncertainties and decision frameworks. In practice, climate adaptation assessments do not only have to deal with uncertainties, but also with time and budget constraints. It might often not be possible to employ all possible methods to deal with all the uncertainties inherent in the assessment. Therefore, it is necessary to prioritise uncertainties and the work needed to assess or reduce them. This can be done in the following three steps:

- Identify and characterise sources of uncertainty;
- Assess (weigh, appraise, and prioritise) sources of uncertainties;
- Select and apply methods for dealing with uncertainties.

In all these steps, the uncertainty typology (see Sect. 2.3) can be used to support the process. Subsequent communication of the results to policy-makers will be discussed in the following paragraphs.

Firstly, the different sources of uncertainty need to be identified. It is likely that a long list of uncertainty sources will be generated and this can be done using two different approaches:

- By analysing each step of the climate assessment at hand and subsequently characterising each source according to the typology, and
- By considering each possible type from the uncertainty typology and discussing where in the assessment this type of uncertainty may occur.

Reasoning from both angles may help to minimise the chance that a source is overlooked. The resulting list of uncertainties can be further characterised using the uncertainty assessment (Sect. 2.3).

Secondly, the relative importance of each uncertainty element can be weighted based on its potential impact on the outcome of the climate assessment in question. Where some form of quantification is possible, the relative importance can be assessed by means of sensitivity analysis. However, for many sources of uncertainty, such quantification is not feasible. In such a case, the relative importance can and should be assessed using expert judgement to consider the importance as being either of crucial, average, medium or low importance. Results from individual experts can be combined to arrive at a group ranking of the items on the list of uncertainties. Arguments used by the experts to defend their ranking need to be documented and special attention should be given to reasons for any substantial disagreement on the importance of a particular uncertainty source.

Thirdly, after the weighing, appraising, and prioritisation, suitable tools can be selected for further analysis of the key uncertainties. Each uncertainty type may require a different method to address it, and to gauge its impact on decision-making. The uncertainty tool catalogue described in Sect. 2.3.4 provides guidance for selecting appropriate methods that match the characterisation of the uncertainty in the typology.

It may, however, not be possible to correctly identify, characterise and prioritise all sources of uncertainty at the beginning of an assessment. The typology may thus need to be reassessed throughout the project. New sources of uncertainty may be added or their weights may be adjusted. The uncertainty typology should therefore be used interactively throughout the study. As such, it also provides a framework for keeping track of all sources of uncertainty, so that those identified early in the project – especially if not immediately quantifiable – are not forgotten at the end of the study when results are reported.

2.7 Cases, Types of Uncertainty, and Methods as Used in Chap. 4

The aforementioned uncertainty assessment methods can be recognised in the various case studies described in Chap. 4. The overview displayed in Table 2.3 gives specific information on every case study.

2.8 Communicating Uncertainty Assessment to Policy-Makers and Decision-Makers

Most policy-makers and decision-makers will feel more comfortable when making decisions based on single, undisputed numbers with small uncertainty ranges, than on ambiguous or controversial estimates and scenario analyses. Unfortunately, however, complex processes cannot often be described this way. There again, giving

Table 2.3 Chapter 4 case study overview

Case studies	Level of Uncertainty	Methods used
Water Supply Management in Portugal (4.2.1)	Scenario	Scenario analysis ("surprise-free") Expert elicitation Sensitivity analysis Stakeholder involvement Extended peer review (review by stakeholders)
UK Climate Change Risk Assessment (4.2.2)	Statistical Scenario	Scenario analysis ("surprise-free") Expert elicitation Sensitivity analysis Bayesian methods NUSAP/Pedigree analysis Stakeholder involvement Quality assurance/Quality checklists Extended peer review (review by stakeholders)
Water Resources Management in England and Wales (4.2.3)	Statistical	Monte Carlo Probabilistic multi model ensemble
Water Supply in Hungary (4.2.4)	Scenario	Expert elicitation Sensitivity analysis Probabilistic multi model ensemble Fuzzy set/imprecise probabilities Stakeholder involvement
Climate Change and Health in The Netherlands (4.2.5)	Scenario Recognised ignorance	Expert elicitation Stakeholder involvement
Flood Risk in Ireland (4.2.6)	Scenario Recognised ignorance	Sensitivity analysis Wild cards/ Surprise scenarios
Coastal Flooding and Erosion in South West France (4.2.7)	Scenario Recognised ignorance	Expert elicitation Stakeholder involvement
Québec Hydro-Electric Power (4.2.8)	Scenario	Scenario analysis ("surprise-free") Expert elicitation Sensitivity analysis Probabilistic multi model ensemble Stakeholder involvement
Austrian Federal Railways (4.2.9)	Scenario	Expert elicitation Sensitivity analysis Bayesian methods Stakeholder involvement
Dresden Public Transport (4.2.10)	Scenario Recognised ignorance	Scenario analysis ("surprise-free") Expert elicitation Sensitivity analysis Fuzzy sets/imprecise probabilities Stakeholder involvement Wild cards/ Surprise scenarios Fuzzy cognitive mapping

(continued)

Table 2.3 (continued)

Case studies	Level of Uncertainty	Methods used
Hutt River Flood Management (4.2.11)	Statistical Scenario	Scenario analysis ("surprise-free") Sensitivity analysis Probabilistic multi model ensemble Stakeholder involvement
Communication of Large Numbers of Climate Scenarios in Dutch Climate Adaptation Workshops (4.2.12)	Scenario	Scenario analysis ("surprise-free") Expert elicitation Sensitivity analysis Stakeholder involvement

policy-makers a lengthy report listing all the possible uncertainties will not necessarily lead to informed policy-making either.

Scientists can help policy-makers (and their respective target groups like shareholders and the general public) by assessing which uncertainties are most relevant for the policy decisions concerned. They can identify policy options that are robust given these uncertainties. If no single best policy option for all scenarios can be determined, all reasonable options can be discussed in a democratic, participatory process including scientists, stakeholders, policy makers and politicians (Pielke et al. 2007). As the communication needs of all these parties can vary greatly, a single mode of risk communication is rarely sufficient.

Uncertainties can be communicated linguistically, numerically, or graphically. Confidence intervals can be provided reflecting uncertainty in parameters and input data. For uncertainties that cannot be expressed in statistical intervals, other characterisations of likelihood can be used. Risbey et al. (2005) have proposed expressions for different levels of precision, ranging from full well defended probability density functions, to percentile bounds, first order estimates, expected signs or trends, ambiguous signs or trends and, finally, effective ignorance. Additionally, if policy recommendations are made, the strength of these recommendations and the quality of the underlying evidence can be expressed using qualitative grading (Atkins et al. 2004; Guyatt et al. 2008).

In order not to overwhelm the user of the assessment results with uncertainties, the concept of progressive disclosure of information can be employed (Wardekker et al. 2008; Kloprogge et al. 2007). This involves tailoring the information about uncertainty to the target audience. In a press release or a project summary, for example, the uncertainties that are most relevant to the final policy decisions need to be described, without any technical details. This way, a policy-maker using the results of a climate assessment will not be directly confronted with a typology of all uncertainties, but will be provided with the information needed to properly interpret the results. The main assessment or background report may subsequently contain more detailed information, with emphasis on the nature, extent and sources of uncertainties. Ideally, it presents all methods, assumptions, parameters and input data, thereby providing maximum transparency of the assessment approach.

2.9 Conclusions

In this chapter we have examined various aspects of dealing with uncertainty in support of decision-making on climate adaptation. To be effective, adaptation should ideally be part of any urban and rural economic development policy and related sectoral plans and budgets. Vulnerability to climate change will be greatly affected by the development and implementation of behavioural, technical, and spatial adaptation strategies and policies. Uncertainty assessment and dealing with uncertainty are integral parts of establishing and implementing targeted climate adaptation policies.

The uncertainty assessment and dealing with uncertainty in adaptation policy can be dealt with in the following ways:

- **Top-down**, prediction-oriented approaches which are strong in statistical uncertainty and can reasonably cope with scenario uncertainty, but cannot handle ignorance.
- Resilient and robust types of **bottom-up** approaches which are strong in coping with recognised ignorance and surprises.

Without knowing too much of the magnitude and nature of climate change impacts, we can still formulate reasonable policies to make the system less prone to possible changes. An essential first step in the selection of an appropriate decision-making framework and methods for uncertainty analysis needs to be based on the policy-relevance of each of the three levels of uncertainty, along with a judgment of their relative importance.

Different strategies and approaches to uncertainty require different scientific methods for assessment. The top-down approaches require probabilistic estimates and (surprise-free) scenarios, such as Bayesian methods and Monte Carlo analysis. The bottom-up approaches use qualitative uncertainty methods such as the NUSAP approach. They also use participatory knowledge production and knowledge assessment, wild cards and surprise scenarios.

Different approaches are available for dealing with uncertainty in adaptation policy. For example, case 4.2.5 shows how resilience can be used for climate adaptation in urban areas in the face of all types of uncertainty, but the effectiveness and efficiency is very difficult to assess in quantitative terms. Predict and control may be appropriate in some management situations while adaptive/resilience-oriented approaches are useful in others. For example, resilience is highly suitable for tailoring bottom-up type of adaptation to the local situation, while the more rigid prediction-oriented approaches is sometimes used by top-down oriented approaches by national and regional governments.

Other factors also influence the usefulness of various strategies: the relevance of the expected impacts; the expected encroachment on society; and, extensiveness of required interventions. For example, we need to ask ourselves whether an approach can be easily implemented in an existing situation, or whether we would need rigorous reforms, redevelopments, or changes in the way we 'do things', and what the costs and co-benefits of actual options would be.

This is demonstrated in case 4.2.5 where precautionary measures deal well with ignorance but can involve high costs and potential side-effects; such approaches are advised for impacts that are both highly uncertain and highly relevant.

For possible climate-related impacts characterised by ignorance, the results of a climate change and health study could be extrapolated towards a more general view (Wardekker et al. 2012; this view is also visualised in a scheme described in the Dutch case study in Sect. 4.2.5):

- Adaptation policies that focus on enhancing the system's and society's capability of dealing with possible future changes, uncertainties and surprises (e.g. through resilience, flexibility, and adaptive capacity) seem most appropriate.
- For climate-related effects for which rough risk estimates are available, 'robust measures are recommended.
- For effects with limited societal or policy relevance, 'no-regret' measures are recommended.
- For highly policy-relevant climate effects, precautionary measures can be considered. However, for such options, it would be advisable to assess the risks of over-investment to avoid excessive costs and to ensure their flexibility.

We advise assessing the availability of 'no-regret' adaptation options as well as the adaptation options that have co-benefits with other policy issues. For quantifiable effects it seems useful to combine system-enhancement with approaches such as 'robust decision-making'. Knowledge gaps on the effectiveness of adaptation options will likely limit adaptation to a qualitative/ semi-quantitative exploration. An exploration of uncertainty typology could contribute to policy/political discussions on the preferred ambition level of adaptation strategies, also considering the range of potential impacts.

There is a growing feeling that a sort of *'dynamic and incremental adaptive strategy'*, taking various sorts and levels of uncertainties into account, is a very promising targeted policy approach, especially for new and ambiguous risks. Analysing and characterising uncertainty by means of a specific typology can be a useful approach for the selection and prioritisation of preferred adaptation policies to reduce future climate related risks. It can also help policy-makers and practitioners to make more educated decisions.

This book will help scientists, decision-makers and policy-makers deal with uncertainty and will show how others, in their specific adaptation cases, have tackled this issue.

References

Atkins, D., D. Best, P.A. Briss, M. Eccles, Y. Falck-Ytter, S. Flottorp, G.H. Guyatt, R.T. Harbour, M.C. Haugh, D. Henry, S. Hill, R. Jaeschke, G. Leng, A. Liberati, N. Magrini, J. Mason, P. Middleton, J. Mrukowicz, D. O'Connell, A.D. Oxman, B. Phillips, H.J. Schunemann, T.T. Edejer, H. Varonen, G.E. Vist, J.W. Williams Jr., and S. Zaza. 2004. Grading quality of evidence and strength of recommendations. *BMJ* 328: 1490.

Briggs, David. 2008. A framework for integrated environmental health impact assessment of systemic risks. *Environmental Health* 7: 61. doi:10.1186/1476-069X-7-61.

Dessai, Suraje 2005. Robust adaptation decisions amid climate change uncertainties. PhD Thesis. University of East Anglia, Norwich.

Dessai, Suraje, and J.P. van der Sluijs. 2007. Uncertainty and climate change adaptation: A scoping study. http://www.nusap.net/downloads/reports/ucca_scoping_study.pdf.

EEA. 2012. Climate change, impacts and vulnerability in Europe 2012. http://www.eea.europa.eu/publications/climate-impacts-and-vulnerability-2012.

EEA. 2012a. Environmental indicator report 2012 – ecosystem resilience and resource efficiency in a green economy in Europe. http://www.eea.europa.eu/publications/environmental-indicator-report-2012/environmental-indicator-report-2012-ecosystem.

Giorgi, Filippo. 2005. Climate change prediction. *Climatic Change* 73: 239–265.

Guyatt, G.H., A.D. Oxman, G.E. Vist, R. Kunz, Y. Falck-Ytter, P. Alonso-Coello, and H.J. Schunemann. 2008. GRADE: An emerging consensus on rating quality of evidence and strength of recommendations. *British Medical Journal* 336: 924–926.

Henderson-Sellers, A. 1993. An antipodean climate of uncertainty. *Climatic Change* 25: 203–224.

IPCC. 2005. Guidance notes for lead authors of the IPCC Fourth Assessment Report on addressing uncertainties. Intergovernmental Panel on Climate Chang (IPCC). http://www.ipcc.ch/pdf/supporting-material/uncertainty-guidance-note_ar4.pdf.

IPCC. 2007a. Climate change 2007: The physical science basis. Intergovernmental Panel on Climate Change (IPCC). http://www.ipcc.ch/publications_and_data/publications_ipcc_fourth_assessment_report_wg1_report_the_physical_science_basis.htm.

IPCC. 2007b. Climate change 2007: Impacts, adaptation and vulnerability. Intergovernmental Panel on Climate Change (IPCC). http://www.ipcc.ch/publications_and_data/publications_ipcc_fourth_assessment_report_wg2_report_impacts_adaptation_and_vulnerability.htm.

Kloprogge, Penny, J.P. van der Sluijs, and J.A. Wardekker 2007. Uncertainty communication: Issues and good practice. http://www.nusap.net/guidance.

Knol, Anne, A.C. Petersen, J.P. van der Sluijs, and E. Lebret. 2009. Dealing with uncertainties in environmental burden of disease assessment. *Environmental Health* 8: 21. doi:10.1186/1476-069X-8-21.

PBL. 2009. News in climate science and exploring boundaries: A policy brief on developments since the IPCC AR4 report in 2007. Netherlands Environmental Assessment Agency (PBL). http://www.pbl.nl/en/publications/2009/News-in-climate-science-and-exploring-boundaries.

Pielke Jr., R., G. Prins, S. Rayner, and D. Sarewitz. 2007. Lifting the taboo on adaptation. *Nature* 445: 597–598.

Ribeiro, M., C. Losenno, T. Dworak, E. Massey, R. Swart, M. Benzie, and C. Laaser. 2009. Design of guidelines for the elaboration of regional climate change adaptations strategies. Study for European Commission, Ecologic institute, Berlin.

Risbey, J., J. van der Sluijs, P. Kloprogge, J. Ravetz, S. Funtowicz, and S. Corral Quintana. 2005. Application of a checklist for quality assistance in environmental modelling to an energy model. *Environmental Modeling & Assessment* 10(1): 63–79.

Runhaar, Hens, H. Mees, A. Wardekker, J. van der Sluijs, and P. Driessen. 2012. Adaptation to climate change-related risks in Dutch urban areas: Stimuli and barriers. *Regional Environmental Change*. doi:10.1007/s10113-012-0292-7.

Schneider, S. 1983. CO2, Climate and society: a brief overview. In *Social science research and climate change: An interdisciplinary appraisal*, ed. R.S. Chen et al., 9–15. Boston: D. Reidel Publishing Company.

Smith, Barry, I. Burton, R.J.T. Klein, and J. Wandel. 2000. An anatomy of adaptation to climate change and variability. *Climatic Change* 45: 223–251.

Utrecht University/RIVM. 2004. *RIVM/MNP guidance for uncertainty assessment and communication: Tool catalogue for uncertainty assessment*, ed. J.P. van der Sluijs, P.H.M. Janssen, A.C. Petersen, P. Kloprogge, J.S. Risbey, W. Tuinstra, and J.R. Ravetz. Utrecht/Bilthoven: Utrecht University.

Van der Sluijs, Jeroen, J.S. Risbey, P. Kloprogge, J.R. Ravetz, S.O. Funtowicz, S. Corral Quintana, Â. Guimaraēs Pereira, B. De Marchi, A.C. Petersen, P.H.M. Janssen, R. Hoppe, and S.W.F. Huijs. 2003. "RIVM/MNP guidance for uncertainty assessment and communication: Detailed guidance." http://www.nusap.net/guidance.

Walker, W.E., P. Harremoës, J. Rotmans, J.P. van der Sluijs, M.B.A. van Asselt, P. Janssen, and M.P. Krayer von Krauss. 2003. Defining uncertainty: A conceptual basis for uncertainty management in model-based decision support. *Integrated Assessment* 4(1): 5–17.

Wardekker, Arjan, J.P. van der Sluijs, P.H.M. Janssen, P. Kloprogge, and A.C. Petersen. 2008. Uncertainty communication in environmental assessments: Views from the Dutch science-policy interface. *Environmental Science & Policy* 11(7): 627–641.

Wardekker, A., A. de Jong, J.M. Knoop, and J.P. van der Sluijs. 2010. Operationalising a resilience approach to adapting an urban delta to uncertain climate changes. *Technological Forecasting and Social Change* 77(6): 987–998.

Wardekker, Arjan, A. de Jong, L. van Bree, W.C. Turkenburg, and J.P. van der Sluijs. 2012. Health risks of climate change: An assessment of uncertainties and its implications for adaptation policies. *Environmental Health* 11: 67. doi:10.1186/1476-069X-11-67.

Willows, R.I. and R.K. Connell, ed. 2003. Climate adaptation: Risk, uncertainty and decision-making. UKCIP Technical Report. UKCIP, Oxford.

Chapter 3
How Is Uncertainty Addressed in the Knowledge Base for National Adaptation Planning?

Hans-Martin Füssel and Mikael Hildén

Key Messages

Fourteen European countries have provided information on the consideration of uncertainty in their knowledge base for adaptation planning, and there are substantial differences across countries and jurisdictions. Some key features are as follows:

- Almost all national-level climate change projections consider uncertainties related to emission scenarios, global climate models and downscaling methods.
- Many countries have established web portals that provide access to climate projections; their functionality and the presentation of uncertainty vary widely across them.
- Only a few countries have developed non-climatic (e.g. socio-economic, demographic and environmental) scenarios for use in climate change impact, vulnerability and risk assessments.

(continued)

All rights reserved
No part of this chapter may be reproduced in any form or by any means electronic or mechanical, including photocopying, recording or by any information storage retrieval system, without a prior permission in writing. For permission, translation or reproduction rights please contact EEA (copyrights@eea.europa.eu)

H.-M. Füssel (✉)
Air and Climate Change Programme, European Environment Agency,
Kongens Nytorv 6, 1050 Copenhagen K, Denmark
e-mail: martin.fuessel@eea.europa.eu

M. Hildén
Climate Change Programme, SYKE, Finnish Environment Institute, Mechelininkatu 34a,
PO Box 140, FI-00251 Helsinki, Finland
e-mail: mikael.hilden@ymparisto.fi

(continued)

- All countries have conducted climate impact, vulnerability or risk assessments. The consideration of uncertainty within these varies widely, from a generic qualitative discussion to a probabilistic assessment based on a comprehensive modelling exercise.
- As adaptation activities expand, an increasing demand for more spatially and temporally detailed and varied climate scenarios brings uncertainties to the forefront.
- Most countries have developed guidance material for decision-makers concerned with adaptation. Such guidelines generally explain key sources of uncertainty in climate and climate impact projections but only few guidelines provide practical guidance on adaptation decision-making under uncertainty.
- Substantial efforts are needed to improve the appreciation of uncertainties in climate and climate impact projections by decision-makers and the public at large.

Dynamic interactive tools in web portals can be an important part of the tool box for those who are confronted with adapting to climate change. In addition, targeted guidance is needed that explains the relevance of key uncertainties and how they can be addressed by appropriate adaptation strategies in a specific adaptation context.

3.1 Introduction

In this chapter we provide an overview of national climate change adaptation planning in Europe with a special focus on the consideration and communication of uncertainties. This provides a context for the consideration of case studies in Chap. 4, which presents 12 adaptation case studies from 10 countries. The link between the national level information presented in this chapter and the case studies for those 6 countries covered in both chapters is briefly discussed in Sect. 3.3.

The chapter is mostly descriptive, highlighting large differences across countries in the information base available to decision-makers concerned with adaptation. It also shows that those countries which are more advanced in the development of adaptation strategies generally pay more attention to the assessment and communication of key uncertainties and to their consideration in policy development. This finding is relevant for countries that are developing or updating their knowledge base for adaptation. In this context, examples from more advanced countries can serve as an inspiration to other countries.

Section 3.2 presents a brief review of national adaptation strategies and action plans. This review is based on information collected by the European Environment Agency (EEA) through the European Climate Adaptation Platform (Climate-ADAPT[1])

---

[1] http://climate-adapt.eea.europa.eu

complemented by two independent scientific studies (see Table 3.1 for details). Section 3.3 reviews the consideration of uncertainties in key information sources for adaptation (climate projections, non-climatic scenarios, climate impact projections and guidance material). This review covers those 14 EEA member countries that have provided pertinent information to the EEA through a questionnaire (see Sects. 3.2 and 3.3 for details).

3.2 Overview of National Adaptation Activities

Most countries in Europe have begun to respond to the impacts of climate change. This is evidenced in:

- The undertaking of research projects related to climate impacts, vulnerability and adaptation,
- The development of climate projections,
- The preparation of climate change impact, vulnerability and risk (CCIV) assessments,
- The increasing availability of web portals related to climate change adaptation, and
- The development of national adaptation strategies and/or action plans.

Adaptation activities differ considerably across countries. This is due to a number of factors, including the following (see also EEA 2013):

- Current and projected future exposure of systems and assets at risk to climatic hazards (e.g. proportion of the population living in coastal zones),
- Existing governance arrangements for climate-sensitive sectors,
- Awareness among the different categories of stakeholders, and
- Available financial and human resources.

There are also considerable differences in the extent of adaptation activities across sectors as well as differences in earmarking certain activities as adaptation. Comprehensive information on the state of adaptation in Europe at European, national, and subnational levels is provided in the recent EEA report *Adaptation in Europe* (EEA 2013) and in Climate-ADAPT. Additional information on national and regional adaptation research efforts is available in the CIRCLE-2 Climate Adaptation INFOBASE.[2]

Table 3.1 provides a summary of national-level adaptation efforts across 28 European countries (all EU member states except for Croatia and Luxemburg, plus Norway and Switzerland, which are EEA member countries) based on a number of sources.[3] The 14 countries marked in grey in the left-most column are those included

---

[2] http://infobase.circle-era.eu

[3] The table includes information from those 27 EEA member countries that have provided information on the country pages in Climate-ADAPT at the end of 2012. The EEA member countries include all EU Member States and additionally Iceland, Liechtenstein, Norway, Switzerland and Turkey.

Table 3.1 Overview of national-level adaptation activities

Country	Stage of selected national activities			Advancement of adaptation		Uncertainty communication in NAS
	CCIV	NAS	NAAP	Policy cycle	Uncertainty	Total score
	0: no activity; 1: in preparation; 2: finalized/adopted			1: assessing risks; 2: identifying options; 3: assessing options; 4: implementation; 5: monitoring and evaluation	1: not mentioned; 2: presented as unreliability; 3: hidden or presented as barrier to adaptation; 4: embracing	0: lowest score; 2: highest score
AT - Austria*	2	2	2	3	3	
BE - Belgium	1+2	2	1+2			1
BG - Bulgaria	1	1	1+2			
CH - Switzerland	2	2	1			
CY - Cyprus	1	1	1			
CZ - Czech Republic	1+2	1	0			
DE - Germany*	1+2	2	2			1.75
DK - Denmark	2	2	2			1
EE - Estonia	1	1	1			
ES - Spain	2	2	2	4	3	
FI - Finland	2	2	2	4	3	2
FR - France*	1+2	2	1+2	3	3	1.5
GR - Greece	1	1	1			
HU - Hungary*	1+2	2	1+2			0.75
IE - Ireland*	1	2	1			
IT - Italy	1	1	1	1	1	
LT - Lithuania	2	2	1+2			
LV - Latvia	1	1	1			
MT - Malta	0	2	0			
NL - Netherlands*	2	2	2			0
NO - Norway	2	1	2			
PL - Poland	1	1	1+2	1	1	
PT - Portugal*	2	2	1			
RO - Romania	2	1	0	2	2	
SE - Sweden	2	2[a]	2			
SI - Slovenia	1	1	1			
SK - Slovakia	1	1	1			
UK - United Kingdom*	2	2	1+2	4	4	
Status	March 2013			2010		2012
Source	EEA (2013, Table 3.1), based on Climate ADAPT			Hanger et al. (2013), based on Pfenninger et al. (2010)		Lorenz et al. (2013)

Countries marked in grey in the left-most column (and with numerical scores in bold face) are included in the detailed analysis in the following section

The traffic-light colours (green, yellow and red) illustrate the numerical values to aid visual comparison

Blank fields in the three right-most columns indicate that a country was not included in the underlying study

Countries marked by an asterisk (*) are represented by one or more case studies in Chap. 4

CCIV climate change impact, vulnerability and risk assessment, *NAS* National Adaptation Strategy, *NAAP* National Adaptation Action Plan

[a]Sweden does not have a specific document called National Adaptation Strategy. Instead Sweden has a set of delegated tasks to national and regional authorities, to produce information useful in adaptation decisions, to provide knowledge and spread knowledge on adaptation, and to regionally coordinate adaptation

in the analysis in Sect. 3.3 because they have provided sufficient information on uncertainties to the EEA through a questionnaire. These 14 countries include the 3 countries with the highest scores according to Hanger et al. (2013) as well as all but one country considered in Lorenz et al. (2013).

The first three columns (from the left) reflect information provided by EEA member countries to Climate-ADAPT and are summarised in a recent EEA report (EEA 2013).[4] The table shows the status of completed and on-going CCIV assessments[5] as well as the status of National Adaptation Strategies (NAS) and National Adaptation Action Plans (NAAP). A NAS is understood here to be a broad policy document that outlines the direction of action in which a country intends to move in order to adapt to climate change. While a NAS shows some political commitment towards climate change adaptation, it does not always imply that adaptation activities are occurring. NAAPs are more detailed documents giving guidance on specific adaptation actions that are being planned. Out of 28 countries included in this table, 17 countries have finalized a CCIV assessment, with several of them already working on a new one. Sixteen countries have adopted a NAS and 15 a NAAP. In most cases, a comprehensive CCIV assessment precedes the adoption of a NAS or NAAP.

The next two columns summarise an assessment of the advancement of adaptation in general and the treatment of uncertainties specifically for a subset of eight countries from a study by Hanger et al. (2013). The study assessed available policy documents and conducted semi-structured interviews with 30 stakeholders. The advancement of adaptation is assessed according to the policy cycle underlying the Adaptation Support Tool in Climate-ADAPT.[6] The same stages are used in the *Guidelines on developing adaptation strategies* (EC 2013) that were published by the European Commission in connection with the EU Adaptation Strategy. The numerical codes cannot be directly compared across columns as they are taken directly from the underlying studies. Comparison across different sources is facilitated by a standardised colour code, which reveals a general agreement between the stage within the policy cycle and the development of an NAS and/or NAAP.[7]

The study authors identified close links between the stage within the policy cycle and the perception of uncertainties: "*the way uncertainty is perceived seems to change with the progression of adaptation policy-making*" (Hanger et al. 2013, pp. 98–99).

---

[4] No information was available for the EEA member countries Liechtenstein, Luxembourg, Iceland and Turkey. Information for Denmark was updated compared to (EEA 2013) following the adoption of the *Action plan for a climate-proof Denmark* (http://en.klimatilpasning.dk/media/590075/action_plan.pdf).

[5] The terms climate impact, vulnerability and risk assessment, as used in different countries, show substantial overlaps. In the context of this study, no further distinction is made within this group of assessments. For a discussion of the evolution of these kinds of assessments, see Füssel and Klein (2006). For a discussion of the use of the terms vulnerability and risk in the climate change context, see the Glossary and EEA (2012, Section 1.7).

[6] http://climate-adapt.eea.europa.eu/web/guest/adaptation-support-tool/step-1

[7] The most noticeable difference between the two sources is related to Poland. The assessment for Poland in Hanger et al. (2013) is based on Pfenninger et al. (2010) and did not consider more recent information available in Climate-ADAPT.

They conclude that *"the farther ahead countries appear to be in adaptation planning and implementation, the better developed is the science-policy interface and the more refined and specific are both the expressed needs for information and the handling of uncertainty. Policy-makers in these countries simply understand the problem better"* (p. 100).

We note that similarities in the relationship between the availability of relevant information and the stage of adaptation policy were found in the EEA Report *Adaptation in Europe* (EEA 2013). It must be considered that the fact that some countries are ahead in adaptation planning could be *because* the science-policy interface has been more refined. For example in Finland, which produced the first NAS in Europe, the whole process started from research activities that were rapidly adopted and transformed into policy documents by the administration and policy-makers.

An independent desk study analysed how uncertainties were represented in the NAS of seven European countries and of three devolved regions of the United Kingdom (Lorenz et al. 2013). The final (right-most) column presents the summary score for the seven countries. Considering that only two countries were included in both studies represented in the two right-most columns, it is not possible to compare the assessments of how uncertainty is addressed between the two studies.[8]

The EEA has led a survey, described more fully in Sect. 3.3, which provides information that is complementary to Lorenz et al. (2013). The restriction to NAS in the Lorenz et al. study provides a well-defined basis for a cross-country comparison, but it excludes a rich variety of information that can be highly relevant for adaptation decision-makers in the country. In contrast, the EEA survey assesses the consideration of uncertainties in the larger knowledge base available for adaptation decision-makers.

3.3 Consideration of Uncertainty in the Knowledge Base for Adaptation

In this section we focus on key information sources intended to support adaptation to climate change in Europe and the way they consider uncertainty. This review encompasses publications and websites dealing with climate change and climate impact scenarios and documents providing guidance for the use of these scenarios in adaptation decision-making. These information sources cover several of the nine essential components for adaptation implementation by governments identified by Smith et al. (2009).

The planning and implementation of activities to adapt to future climate change face substantial uncertainties related to the future development of the climate

---

[8] The very low score for the Netherlands in Lorenz et al. (2013) is due to the fact that this study assessed the National Programme on Climate Adaptation and Spatial Planning from 2007 rather than the more recent Delta Programme.

system and society. Uncertainties generally increase from global emission scenarios through changes in radiative forcing, the global temperature response and changes in regional climate parameters to the range of possible regional impacts (Wilby and Dessai 2010). Uncertainties related to future changes in societal factors (including demography, economy, technology and governance) and in environmental factors (including land use) are crucial for determining social impacts of climate change and adaptation needs.

Numerous typologies have been developed to distinguish different sources and types of uncertainty relevant for adaptation planning (see also Sect. 2.3). A fundamental distinction of sources of uncertainty relevant for future projections is between decision uncertainty (e.g., related to human decisions that determine future greenhouse gases and aerosol particle emissions), natural variability (e.g., related to the internal variability of the climate system), and scientific uncertainty (e.g., related to data gaps, incomplete understanding or insufficient computing power of climate and climate impact models). For further information, see Chap. 2.

For the purpose of this assessment, the EEA has developed a questionnaire that addresses three broad aspects of uncertainty and adaptation:

- The provision of quantitative scenarios (further distinguished into climate projections, non-climatic projections, and climate impact/vulnerability/risk assessments),
- The provision of guidance material, and
- Legal requirements.

A first set of responses was collected by the EEA through the Interest Group on 'Climate Change and Adaptation' of the Network of European Environmental Protection Agencies (EPA IG Adaptation). An updated version of the questionnaire was later sent to the National References Centres (NRCs) on Climate Change Impact, Vulnerability and Adaptation of those EEA member countries from which no response was received through the EPA IG on adaptation. NRCs are typically either the Ministry in charge of Environment and Climate or the Environmental Agency in an EEA member country. The information reported through the questionnaire has been complemented by us based on various publicly available information sources.

Responses from 14 countries are included in this analysis (see the grey shading in Table 3.1). These are from countries that provided, as a minimum, links to publicly available climate change projections.[9]

---

[9] Further responses were received from Croatia, Lithuania and Slovenia. Croatia and Slovenia were not included in this analysis because their responses contained very limited information on climate projections and the consideration of uncertainties. Lithuania was not included because publicly available information on climate and climate impact projections was largely restricted to National Communications under the United Nations Framework Convention on Climate Change (UNFCCC). Note that information for "Belgium" was reported separately for the Flemish and the Walloon region, and some information is only available for one of these regions. One member of the EPA IG on Adaptation provided a response for the Basque Autonomous Region in Spain. This response was excluded considering that comprehensive information for Spain was available separately.

3.3.1 Sources of Uncertainty in Climate Change Projections

Uncertainty about future climate change is a key consideration for planning adaptation to climate change. In Chap. 2 we discussed key sources of uncertainty along the chain from global climate projections to regional climate change impacts and adaptation needs. Table 3.2a gives an overview of the sources of uncertainty (emissions scenarios, global climate models [GCMs] and regional climate models [RCMs]) that were considered in climate change projections provided or authorised by national governments in the 14 countries in this survey.[10]

Status

The column titled "Status" reveals that the use and official status of climate projections varies widely across countries. In Switzerland, use of an optimistic and a pessimistic climate projection is mandatory for federal offices in the context of the development of the Swiss action plan. The UKCP09 projections for the United Kingdom also have a strong status as their use is recommended in the preparation of climate change risk assessments as required by the Climate Change Act 2008. In several other countries, the climate projections reviewed here are mentioned in official documents or are the de facto standard due to the absence of alternative projections of comparable quality.

Time Horizon

Most climate projections included in Table 3.2a cover the period until 2100, which corresponds to the time horizon of Special Report on Emissions Scenarios (SRES) emissions scenarios (Nakicenovic and Swart 2000) and of the ENSEMBLES project (see below). The current *reclip:century* project scenarios for Austria have a time horizon until 2050, which will be extended to 2100 in phase 2 of the project. The KNMI'06 climate scenarios for the Netherlands extend until 2050, but the scenarios used in the *Klimaateffectatlas* (Climate Impact Atlas) and the Dutch Delta Programme include projections of sea-level rise and water-related climate variables until 2100 (Delta Programme 2011).

---

[10] The table contains two different sets of climate scenarios for Germany, denoted as Deutscher Klimaatlas (German climate atlas, by the German Weather Service) and Regionaler Klimaatlas Deutschland (Regional climate atlas Germany, by the Regional Climate Offices of the Helmholtz Association). Another set of climate projections for Germany is being provided on the Kompass website of the Umweltbundesamt (Federal Environment Agency). The Kompass projections are not considered here as they are older than the two projections included in Table 2. Spain has published regional climate change scenarios in 2009 and is currently compiling new scenarios from different sources. The Netherlands have also published two sets of climate projections.

Table 3.2a Climate change projections: status and consideration of uncertainties

Country[a]	Name of projection (or portal)[b]	Date	Web link[c]	Status	Time horizon	No. of emission scenarios used	No. of GCMs used	No. of RCMs used
AT	reclip:century	2011	http://tiny.cc/ccp-at	1	2050[d]	2	2	2
BE	Regional projections (Walloon region)	2011	http://tiny.cc/ccp-be1	2	2100[e]	1[d]	3	3**
BE	CCI-HYDR & INBO (Flemish region)	2009	http://tiny.cc/ccp-be2 http://tiny.cc/ccp-be3	2	2100	3[d]	3	3**
CH	CH2011	2011	http://tiny.cc/ccp-ch1 http://tiny.cc/ccp-ch2	4[f]	2100	3	4*	9
CZ	Projekt VaV 2007-2011	2011	http://tiny.cc/ccp-cz1 http://tiny.cc/ccp-cz2	1	2100	1[d]	1[d]	1[d]
DE	Deutscher Klimaatlas	2011	http://tiny.cc/ccp-de1	2	2100	5	4*	11
DE	Regionaler Klimaatlas	?	http://tiny.cc/ccp-de2	1	2100	4	3*	3
ES	Escenarios regionalizados de cambio climático	2009	http://tiny.cc/ccp-es1	1	2100	2	3	9**
ES	PNACC 2012	2013	http://tiny.cc/ccp-es2 http://tiny.cc/ccp-es3 http://tiny.cc/ccp-es4 http://tiny.cc/ccp-es5	3[g]	2100	3	3	3**
FI	ACCLIM	2009	http://tiny.cc/ccp-fi1 http://tiny.cc/ccp-fi2 http://tiny.cc/ccp-fi3	2[h]	2100	3	19*	9
FR	Climat de la France au XXIe siècle	2012	http://tiny.cc/ccp-fr1 http://tiny.cc/ccp-fr2	2	2100	3	3	2**
HU	OMSZ 2008[i]	2008	http://tiny.cc/ccp-hu	1	2100	1[d]	2[d]	2[d]
IE	C4I	2008	http://tiny.cc/ccp-ie	1	2100	4	5	2**
NL	KNMI'06	2006, 2009	http://tiny.cc/ccp-nl1 http://tiny.cc/ccp-nl2 http://tiny.cc/ccp-nl3	3[j]	2050[d]	n.a.[d]	5	10
NL	Klimaateffectatlas	2009	http://tiny.cc/ccp-nl4	2	2100	n.a.[d]	Not specified	
NO	Klima i Norge 2100	2009	http://tiny.cc/ccp-no1 http://tiny.cc/ccp-no2	2	2100	3	6	10**
PL	Projekcje klimatu	?	http://tiny.cc/ccp-pl	1	2100	1	4	7
UK	UKCP09	2009	http://tiny.cc/ccp-uk1 http://tiny.cc/ccp-uk2	3[k]	2100	3	1[d]*	1**

Status: 1: No official status; 2: Reference in official documents/de facto standard; 3: Use officially recommended; 4: Use officially required

No. of GCMs used: An asterisk (*) denotes that a perturbed physics ensemble was produced by at least one of the GCMs

No. of RCMs used: A double asterisk (**) denotes that empirical-statistical downscaling models were applied in addition to RCMs

[a]See Table 3.1 for abbreviations of countries
[b]Projections highlighted in grey were used in case studies described in Chap. 4
[c]This document uses dynamic short links ("tiny URLs") in order to improve the readability of the web link and to allow for an update if an URL changes. Please report broken links to the first author of this book chapter
[d]See text for details
[e]The text states 2085, which is the central year of the period 2071–2100. For consistency with references to the same period in other projections, this is denoted here as 2100
[f]For the development of the Swiss action plan, the federal offices are to consider an "optimistic" scenario and a "pessimistic" scenario
[g]Scenarios-PNACC 2012 is intended to become the official information platform for regionalised climate change scenarios for Spain
[h]Consideration of uncertainty is implicitly required by water managers and electric utility companies
[i]Not an official name
[j]The *Nationaal Bestuursakkoord Water* provides advice on which of the KNMI'06 climate scenarios to use for a specific application
[k]Use of UKCP09 scenarios (and quantification of uncertainties, where appropriate) is recommended in the preparation of Climate Change Risk Assessments (CCRAs) as required by the Climate Change Act 2008

Table 3.2b Climate change projections: communication of uncertainties

Country[a]	Name of projection (or portal)	Climate variables[b]	Data download	Interactive maps	Uncertainty in maps	Uncertainty in graphs (e.g. time series)
AT	reclip:century	2	✓	✓	Individual simulations	–
BE	Regional projections (Walloon region)	2	–	–	Not applicable, because detailed projections are available on request only	–
	CCI-HYDR & INBO (Flemish region)	4	–	–	Individual simulations	Individual simulations
CH	CH2011	6	✓	–	Multi-model mean	Individual simulations; 3 percentiles (2.5th, median, 97.5th); uncertainty range
CZ	Projekt VaV 2007–2011	4	–	–	Individual simulations; mean; robustness of sign (for ENSEMBLE projections)	Individual simulations; percentiles (quartiles, for ENSEMBLE projections)
DE	Deutscher Klimaatlas	9	–	✓	3 percentiles (15th, median, 85th)	Individual simulations
	Regionaler Klimaatlas	23	–	✓	Individual simulations; uncertainty range (min, max); robustness of sign	–
ES	Escenarios regionalizados de cambio climático	8	–	–	Individual simulations;multi-model mean	Individual simulations; multi-model mean; uncertainty range (±1 standard deviation)
FI	PNACC 2012	3	✓	–	–	5 percentiles (min, 25th, median, 75th, max)
	ACCLIM	7	–	✓	Multi-model mean	Multi-model mean; 2 percentiles (5th and 95th)

FR	Climat de la France au XXIe siècle	21	✓	Individual simulations	2 percentiles (2.5th and 97.5th)
HU	OMSZ 2008	2	—	Individual simulations	—
IE	C4I	12	—	Individual simulations; multi-model mean	—
NL	KNMI'06	4	—	Best guess for each of the 4 scenarios	Uncertainty range (not exactly specified)
	Klimaateffect-atlas	47	✓	Best guess for each of the 4 scenarios	—
NO	Klima i Norge 2100	13	—	Individual simulations; multi-model/scenario mean; 3 percentiles (5th, median, 95th)	Multi-model/scenario mean; 3 percentiles (10th, median, 90th); uncertainty range (±1 standard deviation); individual simulations
PL	Projekcje klimatu	2	—	Multi-model mean; 5 percentiles (minimum, 10th, median, 90th, maximum)	—
UK	UKCP09	9	✓	Separately for 3 emissions scenarios: 3 percentiles (10th, median, 90th); multi-model mean	Multi-model mean; probability/cumulative density function; joint probability plot

[a]See Table 3.1 for abbreviations of countries

[b]This information is only indicative because counting the number of climate variables involves several challenges. First, different portals have different approaches in presenting different statistics (e.g., seasonal information) for the "same" climate variable. Second, some portals also include variables that could be described more appropriately as climate impact variables. Note that projections for some emission scenarios may not be available for all variables

Emissions Scenarios

Most climate projections consider simulations forced by 2–5 different emissions scenarios. The approach applied by the Netherlands differs from those of the other countries. Instead of sampling the forcing uncertainty from different emissions scenarios and the climate response from different climate models separately, four climate projections were produced that capture a large range of the variation of those factors that are considered most relevant for the Dutch climate: change in global temperature and change in circulation patterns. A similar approach was used for the climate projections for the Walloon and Flemish regions of Belgium.

The climate projections for the Czech Republic, Hungary and Poland consider only one emissions scenario (SRES A1B); those for the Czech Republic and Poland are furthermore based on a single projection of an RCM (regional climate model) nested in a GCM (general circulation model, also translated as global climate model). However, the Czech projections have been validated and compared with ensemble-based projections based on the EU projects ENSEMBLES[11] and CECILIA.[12] The "Vahava Report" for Hungary (see Table 3.4) used more comprehensive climate scenarios from the PRUDENCE[13] project that are based on 2 emissions scenarios, 3 GCMs and 18 GCM/RCM combinations.

Climate Models

All but two climate projections are based on a multi-model ensemble of 2–19 different GCMs. Several projections also consider different versions of the same GCM or perturbed-physics ensembles in which alternative variants of a single GCM are created by altering the values of uncertain model parameters (Meehl et al. 2007, Section 10.5.4.2). The UKCP09 probabilistic climate projections were produced in a different way. They are based on a large perturbed-physics ensemble of a single GCM but 12 additional GCMs participating in the Cloud Feedback Model Intercomparison Project (CFMIP[14]) were used in the estimation of structural errors.

All climate projections applied RCMs to downscale the coarse GCM projections to a higher resolution; most of them employed several (up to 11) different RCMs. The UKCP09 projections for the United Kingdom employed only one RCM due to the large number of simulations required for the probabilistic projections. Seven climate projections additionally employed empirical-statistical downscaling methods (ESDMs).

---

[11] http://ensembles-eu.metoffice.com
[12] http://www.cecilia-eu.org
[13] http://prudence.dmi.dk
[14] http://cfmip.metoffice.com

Discussion

While there are notable differences in the national climate change projections covered in this analysis, almost all projections share the following characteristics:

- Consideration of different emissions scenarios (see the note above for the Netherlands and for Belgium),
- Use of different GCMs, and
- Downscaling of GCM outputs by different dynamical and sometimes also statistical models.

As can be seen therefore, almost all of the climate projections address the major sources of uncertainty to some degree. This degree of coherence is not surprising considering that the EU-funded projects PRUDENCE (2001–2004) and in particular ENSEMBLES (2006–2009) have been crucial sources for regionalised climate change projections in many countries.[15] An analysis of how national climate scenarios differ from those developed for the whole Europe would be interesting but is beyond the scope of this chapter.

Six countries included in this uncertainty analysis are also covered by adaptation case studies in Chap. 4:

- Case studies in three of these countries (Austria: case 4.2.9, the Netherlands: cases 4.2.5 and 4.2.12 and United Kingdom: case 4.2.2) applied national-level climate scenarios included in Table 3.2a.
- Case studies from two other countries used tailor-made climate change scenarios at the national scale (Ireland: case 4.2.6) or regional scale (Germany: case 4.2.10).
- The French case study (case 4.2.7) did not specify the specific source of climate projections considered, if any.

The case study for the United Kingdom (case 4.2.2) describes the national-level CCIV assessment but none of the other case studies directly uses information from the national-level CCIV assessment (see Table 3.4).

This observation suggests that the current generation of national-level CCIV assessments generally is not well suited to support concrete adaptation planning. It would be interesting to investigate further whether the gap between the information provided in current national-level CCIV assessments and the information needs of local and regional adaptation actors is primarily related to insufficient detail in science-based projections (which could, in principle, be overcome by improved

---

[15] The latest initiative to generate regional climate change projections based on a multi-model ensemble is the CORDEX (http://cordex.dmi.dk/joomla/index.php) project coordinated by the World Climate Research Programme (WCRP). EURO-CORDEX (http://www.euro-cordex.net/) is the European branch of the CORDEX initiative and will produce ensemble climate simulations based on multiple dynamical and empirical-statistical downscaling models forced by multiple GCMs from the Coupled Model Intercomparison Project Phase 5 (CMIP5).

national-level CCIV assessments) or to the insufficient consideration of the specific decision context (which can only be addressed in local or regional-scale assessments involving relevant stakeholders).

3.3.2 Communication of Uncertainty in Climate Change Projections

The discussion above revealed that almost all climate change projections reviewed here consider the main sources of uncertainty to some degree. We noted in Chap. 2 that projections and their associated uncertainties need to be communicated to climate impact researchers from diverse sectors and/or to decision-makers involved in adaptation and risk reduction. They need to understand the robustness of projections relevant for their activities and decisions. Uncertainty generally increases along the impact chain, but it may be possible to find robust adaptation measures even when impact projections are very uncertain.

The consistent, accurate and understandable communication of uncertainties has been the focus of climate scientists, communication psychologists, and others (Budescu et al. 2009; Moser 2010; Fischhoff 2011; Pidgeon and Fischhoff 2011; Lemos et al. 2012; Rabinovich and Morton 2012). The IPCC has made an unprecedented effort to accurately assess uncertainties and consistently communicate the robustness of specific statements in its assessment reports (Moss and Schneider 2000; IPCC 2005; Mastrandrea et al. 2010). At the same time, decision-makers are not always able to make use of the complex information base due to cognitive, institutional, legal, and other reasons.

A clear conclusion from the pertinent literature is that the communication of climate information with its associated uncertainties needs to be audience-specific. For example, Tang and Dessai (2012) found that the saliency of the (probabilistic) UKCP09 projections was dependent on the scientific competence of its users; furthermore, they claim that *"the use of Bayesian probabilistic projections […] improved the credibility and legitimacy of UKCP09's science but reduced the saliency for decision-making"* (p. 300). A one-size-fit-all approach for the communication of climate projections is unlikely to be successful. This is because of the large differences in the information needs of potential users as well as their ability to comprehend complex, and potentially ambiguous, scientific information. Furthermore, knowledge providers also have different ways of framing and communicating uncertainties, e.g. dependent on their disciplinary background (Swart et al. 2009).

Comprehensiveness

Table 3.2a shows the status of all climate projections and Table 3.2b summarises how their results are presented graphically. The column "Variables" shows that some climate change projections are significantly more comprehensive than others. Some of them provide projections for annual and seasonal temperature and precipitation only, whereas others comprise statistics for dozens of climate variables. A detailed assessment of these differences is beyond the scope of this chapter.

Availability of Data and Maps

Five out of 18 climate change portals enable download of the raw data for use in climate impact research and adaptation planning. Eight portals allow for the interactive creation of maps, although with considerable differences in the specific features. The majority of national climate projections are currently only available as static maps and/or graphs. Evidence from one of the case studies ("Communication of large numbers of climate scenarios in Dutch climate adaptation workshops", case 4.2.12) suggests that the presentation of climate projections through interactive maps is very effective in communicating key aspects of future climate change to decision-makers. Hence, the development of interactive web portals could be an important part of developing and sharing the knowledge base for adaptation.

Uncertainty Communication in Graphs and Maps

There are large differences in the presentation of different sources of uncertainty in maps and graphs. Maps focus on *spatial* variations of *one* climate statistic. Many maps present the results from individual model simulations separately. Some maps show climate statistics, including (ensemble) mean, median, various other percentiles and robustness of sign. In most cases, the statistics were calculated across all GCM/RCM combinations for *one* emission scenario. One exception is the *Regionaler Klimaatlas* (regional climate atlas, Germany) where maps depicting the robustness of projections are based on a multi-model ensemble that comprises *all* emissions scenarios. Similarly, map-based projections for Norway are based on a multi-model ensemble forced by different emissions scenarios. The percentiles used to depict "low" and "high" projections vary widely (e.g. "low" projections are based on the minimum as well as the 2.5th, 5th, 10th and 15th percentile).

Presentations of climate projections in graphs often show time series for one climate variable in a particular region. Others show projections for several regions and/or seasons for one time period. In many cases several individual simulations and/or several statistics (e.g. different percentiles) are shown together. UKCP09 offers the widest variety of map and graph-based presentations. Its probabilistic climate projections are presented, among others, as probability density functions, cumulative density functions and joint probability plots for two climate variables.

Summary on Communication of Uncertainties in Climate Projections

The communication of uncertainties in climate projections differs substantially across countries. In some countries, the only available projections are averages of the most important climate variables provided in reports. Such information may serve some general educational purpose but can be misleading when trying to make specific adaptation decisions involving uncertainties, for example, in the level of flood defence required. In other countries, sophisticated web portals provide access to a wide range of user-defined maps and graphs as well as to the underlying data. Such detailed and sophisticated information can provide support for decisions related to risk management. However, its correct interpretation may require specialists, and a general user may lose the wider picture.

The climate information available in some countries is clearly insufficient to fulfil the information needs of many (potential) users. An improvement of this situation requires a dialogue between information providers and key users and careful consideration of user needs already in the design phase of communication tools for climate projections (e.g. reports and web portals).

Most likely, a tiered set of communication material will be required. In such an approach, highly aggregated projections can support initial coarse vulnerability assessments and provide relevant background information for stakeholders whose activities are only moderately sensitive to climate change. More detailed projections, including quantitative uncertainty assessments, provide further information for stakeholders with more detailed information requirements.

3.3.3 Non-climatic Scenarios

Planned adaptation is driven by projected changes in climate, but, like any long-term planning, anticipated changes in other social, economic, and environmental factors also need to be considered. Some projected changes in non-climatic factors can be considered rather certain (e.g. an increasing share of elderly people in most countries in Europe) whereas others are partly speculative (e.g. technological development or the future role of biomass as an energy carrier).

Table 3.3 summarises the availability of non-climatic scenarios for CCIV assessments. Only Finland, the Netherlands and the UK have developed quantitative scenarios for non-climatic variables specifically for CCIV assessments. The Finnish FINADAPT scenarios comprise several variables related to population, economy and environment that are consistent with 3 out of 4 SRES scenario families. The Dutch WLO and IC11 scenarios comprise 26 variables that also cover energy, transport and agriculture. Within the Dutch Delta programme integrated scenarios have been developed that combine the KNMI06 climate scenarios and the WLO socio-economic scenarios in a coherent way (Deltaprogramma 2011). The UK SES scenarios (from 2001) provide quantitative projections for 12 variables and qualitative projections for further topics from similar topic areas as the Dutch scenarios. Switzerland is currently developing socio-economic scenarios for climate change impact assessment.

The Flemish region of Belgium has published socio-economic scenarios for environmental policy planning, which have been considered in the Flemish Adaptation Plan, and Germany has published land use change scenarios (see Table 3.3 for details).

Table 3.3 Availability of non-climatic scenarios for CCIV assessments

Country[a]	Date	Name	Web link	Comment
BE	2009	Environment Outlook 2030 – Flanders	http://tiny.cc/ncs-be	A single scenario for demography, economic development, employment and energy prices
DE	2012	Trends der Siedlungs-flächenentwick-lung – Status quo und Projektion 2030	http://tiny.cc/ncs_de	Regionalised scenarios for changes in land use
FI	2005	FINADAPT scenarios for the twenty-first century	http://tiny.cc/ncs-fi1	Downscaled scenarios of population, sector-specific GDP, household consumption, nitrogen deposition and land use consistent with 3 out of 4 SRES scenario families
	2007	Assessing the adaptive capacity of the Finnish environment and society under a changing climate: FINADAPT	http://tiny.cc/ncs-fi2	
NL	2006	Welfare, Prosperity and Quality of the Living Environment (WLO)	http://tiny.cc/ncs-nl1	The 4 WLO scenarios comprise 26 variables related to demography, economy, housing, industrial areas, mobility, energy, agriculture and environment. They were re-evaluated in 2010 and they provide the basis for the IC11 scenarios.
	2010	Bestendigheid van de WLO-scenario's	http://tiny.cc/ncs-nl2	
	2011	Socio-economic Scenarios in Climate Assessments (IC11)	http://tiny.cc/ncs-nl3	
UK	2001	Socio-economic scenarios for climate change impact assessment (SES)	http://tiny.cc/ncs-uk	The 4 SES scenarios aligned with the 4 SRES scenario families provide quantitative projections up to 2050 for 12 variables related to economic development, population and land use. Further qualitative scenarios are given for those thematic areas as well as for values and policy, agriculture, water, biodiversity, coastal zone management and built environment. The SES scenarios were critically reviewed in 2009.

[a]See Table 3.1 for abbreviations of countries

However, these socio-economic scenarios are not necessarily consistent with the scenarios underlying the climate change projections, and it is not clear whether they have been used in CCIV assessments. Similar projections may also be available in other countries, but they have not been reported.

In summary, most countries lack readily available long-term scenarios of key non-climatic variables that could be used together with climate scenarios to assess potential climate change impacts.

3.3.4 Climate Impact, Vulnerability, and Risk Assessments

Most decision-makers involved in adapting to climate change are less interested in future changes in climate than in the environmental, social, economic, and health risks (and opportunities) associated with them. CCIV assessments aim to provide such information. Table 3.4 gives an overview of national-level CCIV assessments in the 14 countries covered by our analysis. All 14 countries have published CCIV assessments covering key climate-sensitive sectors and systems, and several countries are currently updating them. For a recent overview of CCIV assessments in 7 European countries, see Steinemann and Füssler (2012).

The multi-sector CCIV assessments shown in the table differ considerably in their method, scope, extent, level of quantification and consideration of uncertainties. Many CCIV assessments comprehensively cover a whole country or region whereas others are restricted to individual sectors or systems. About half of them can be categorised as predominantly quantitative and the other half as predominantly qualitative. Some assessments are literature reviews of existing studies whereas others build on consistent multi-sector modelling exercises. Several assessments present quantitative information on uncertainty derived from different climate projections. However, uncertainty arising from non-climatic projections or from impact models is rarely explicitly considered, which may result in maladaptation. Decision-makers are generally well aware of the main non-climate-related uncertainties relevant for their decisions. However, inclusion of such experience-based knowledge in adaptation decisions may be impaired if CCIV assessments present projected impacts of climate change without consideration of other changes and related uncertainties. Therefore, CCIV assessments should ideally consider multiple plausible scenarios for relevant non-climatic developments. Furthermore, they should either be based on multiple climate impact models or discuss how limitations of a given impact model could affect its results.

The UK Climate Change Risk Assessment (CCRA) stands out in many ways: it is the only legally mandated CCIV assessment; it builds on the most comprehensive climate projections (UKCP09); it is the only probabilistic CCIV assessment, providing the 10th, 50th and 90th percentile of projected impacts; and it is the most comprehensive example, comprising several thousand pages. This assessment is described in case study 4.2.2.

Table 3.4 National climate change impact, vulnerability and risk (CCIV) assessments

Country[a]	Date	Name	Web link	Comment
AT	2010	Klimaänderungsszenarien und Vulnerabilität	http://tiny.cc/civ-at	Qualitative; part of the NAS
BE	2011	L'adaptation au changement climatique en région wallonne (Walloon region)	http://tiny.cc/civ-be1	Mostly qualitative; the reporting of sector-specific impacts distinguishes five levels of probability and five levels of quality of the knowledge base
	2010	Bouwstenen om te komen tot een coherent en efficiënt adaptatieplan voor Vlaanderen (Flemish region)	http://tiny.cc/civ-be2	Mostly qualitative; quantitative impact projections are available from some of the underlying research projects
CH	2007	Climate Change and Switzerland 2050: Expected Impacts on Environment, Society and Economy	http://tiny.cc/civ-ch	Mostly qualitative; uncertainty is discussed qualitatively[b]
CZ	2011	Zpřesnění dosavadních odhadů dopadů klimatické změny v sektorech vodního hospodářství, zemědělství a lesnictví a návrhy adaptačních opatření (in Czech)	http://tiny.cc/civ-cz	Quantitative; restricted to water management, agriculture and forestry
DE	2005	Climate Change in Germany. Vulnerability and Adaptation of climate sensitive Sectors	http://tiny.cc/civ-de1	Quantitative; uncertainty resulting from different emissions scenarios and climate models[c]
	2008	Deutsche Anpassungsstrategie an den Klimawandel	http://tiny.cc/civ-de2	Qualitative; part of the NAS
ES	2005	ECCE – A preliminary General Assessment of the Impacts in Spain Due to the Effects of Climate Change	http://tiny.cc/civ-es	Quantitative; based on a comprehensive review of available studies; uncertainty is addressed differently depending on the underlying study[d]
FI	2012	Miten väistämättömään ilmastonmuutokseen voidaan varautua (ISTO)	http://tiny.cc/civ-fi	Mostly qualitative[e]
FR	2009	Climate change: costs of impacts and lines of adaptation	http://tiny.cc/civ-fr	Quantitative; based on a comprehensive review of available studies and specific assessments; uncertainty is addressed by considering a low and a high emission scenario
HU	2010	Climate Change and Hungary: Mitigating the Hazard and Preparing for the Impacts (The "Vahava" Report)	http://tiny.cc/civ-hu	Qualitative; the underlying reports were not available for further analysis

(continued)

Table 3.4 (continued)

Country[a]	Date	Name	Web link	Comment
IE	2008	Climate Change: Refining the Impacts for Ireland	http://tiny.cc/civ-ie1	Quantitative; many uncertainties are presented quantitatively[f]
	2009	A Summary of the State of Knowledge on Climate Change Impacts for Ireland	http://tiny.cc/civ-ie2	Qualitative; based on a literature review
NL	2012	Effecten van klimaatverandering in Nederland 2012	http://tiny.cc/civ-nl	Quantitative; uncertainties covered by 4 KNMI'06 scenarios
NO	2010	Adapting to a changing climate: Norway's vulnerability and the need to adapt to the impacts of climate change	http://tiny.cc/civ-no	Mostly qualitative; uncertainties are mentioned in the text
PL	2010	Opracowanie wskaźników wrażliwości sektora transportu na zmiany klimatu	http://tiny.cc/civ-pl1	Semi-qualitative; consideration of uncertainties not known; only one sector (transport)
	2012	Strategiczny plan adaptacji dla sektorów i obszarów wrażliwych na zmiany klimatu do roku 2020	http://tiny.cc/civ-pl2	Semi-qualitative; consideration of uncertainties not known (due to inavailability of English translation)
UK	2012	The first UK Climate Change Risk Assessment	http://tiny.cc/civ-uk	Comprehensive; quantitative; probabilistic (results for 10th/50th/90th percentile); legally mandated every 5 years

[a]See Table 3.1 for abbreviations of countries
[b]A follow-up project is currently under development (http://www.wsl.ch/fe/wisoz/projekte/klimarisiken/index_DE)
[c]Currently two projects are working at a consistent and cross-sectorial vulnerability assessment for Germany in support of the German Adaptation Strategy
[d]Between 2008 and 2011, CCIV assessments were conducted for the following sectors and systems: biodiversity, water resources, forests, coasts, desertification, and tourism
[e]New studies are currently being undertaken within the Finnish Research Programme on Climate Change (FICCA, http://www.aka.fi/en-GB/A/Programmes-and-cooperation/Research-programmes/Ongoing/FICCA/)
[f]A new national CCIV assessment has been completed but the results have not been published to date

3.3.5 Guidance for Adaptation Planning Under Uncertainty

Climate projections and CCIV assessments provide crucial information for adaptation planning, but this information is often presented in a way that is difficult to understand for adaptation decision-makers (Lemos et al. 2012). Uncertainties in projections present particular challenges for decision-makers as they may be difficult to comprehend or current decision-making criteria may be based on the use of a single "best" value. Therefore, most adaptation decision-makers need help to make best use of available climate and climate impact projections. This section presents a brief overview how uncertainties in climate and climate impact projections are addressed in written guidance material and web-based tools targeted at adaptation decision-makers. A wider analysis of the available guidance material is beyond the scope of this chapter.

Table 3.5 provides an overview of how uncertainties are addressed in guidance documents and websites for adaptation decision-makers across different countries.[16] Apart from the Netherlands, these guidance documents are only available in the national language. Only four countries (Germany, the Netherlands, Norway and United Kingdom) currently explicitly address climate uncertainties in their guidance material for adaptation decision-makers. Finland has published relevant guidance documents for specific sectors, and Spain is developing a user guide where climate uncertainties are addressed. The most comprehensive effort at assisting public and private adaptation decision-makers has been made in the United Kingdom.

The lack of guidance in some countries is surprising. For example, the CCIV assessment for Ireland provides substantial information on uncertainties in climate and climate impact projections but there are no documents helping adaptation decision-makers to address these uncertainties. In addition, while Austria is relatively advanced in terms of adaptation policy (see Table 3.1) and has included several sources of uncertainties in its national climate change projections (see Table 3.2a), information on addressing uncertainties is very difficult to find on its web site.

We conclude that guidance material for addressing uncertainties in adaptation planning is insufficient in most countries. This is even the case in some countries where climate projections or CCIV assessments consider key uncertainties. This means that in most countries, substantial efforts are needed to improve the appreciation of uncertainties in climate and climate impact projections by decision-makers and the public at large. Until the results of these efforts will become available, the reader will have to rely on the sources mentioned in this chapter and additional material available through contacts at the national and local level. Generic understanding of uncertainties at the European (e.g., Climate-ADAPT) and national level (e.g., UKCIP) can be relevant in any adaptation situation in Europe.

---

[16] When interpreting the information in Table 5, it should be considered that guidance documents can possibly be provided by many institutions. It is thus much more difficult to assemble a complete overview of guidance documents than of national-level climate projections and CCIV assessments.

Table 3.5 Guidance on dealing with uncertainty in climate or climate impact projections

Country[a]	Date	Name	Web link	Further information
AT	2011	Der Zukunft vorgreifen: Klimawandelanpassung und Unsicherheiten	http://tiny.cc/gdu-at	Some information on sources of uncertainties and implications for adaptation planning[b]
DE	2010	Klimalotse (Step 3.1)	http://tiny.cc/gdu-de1	Some recommendations on addressing uncertainties related to emission scenarios, global and regional climate models, and development of society and economy
	2012	Stadtklimalotse	http://tiny.cc/gdu-de2	Recommendations on flexible planning under uncertainties
FI	2012	Hulevesiopas	http://tiny.cc/gdu-fi1	Guidance documents on water management in a future climate
	2012	Energialaskennan testivuodet tulevaisuuden ilmastossa	http://tiny.cc/gdu-fi2	Guidance on future climatic reference conditions for architects
NL	2009	Klimaatschetsboek Nederland (Sect. 2.3)	http://tiny.cc/gdu-nl1	Explanation of sources of uncertainty; simultaneous presentation of results for 4 KNMI06 scenarios
	2009	Socio-economic Scenarios in Climate Assessments	http://tiny.cc/gdu-nl2	Guidance for the combination of socio-economic scenarios with climate scenarios
NO	2009	Klima i Norge 2100 (Chap. 6)	http://tiny.cc/gdu-no1	Explanation of sources of uncertainty in climate projections; very brief discussion on dealing with this uncertainty
	2012	Klimaprojeksjoner og usikkerhet	http://tiny.cc/gdu-no2	Guidance on the consideration of climate uncertainties for municipalities
UK	2013	Climate change: Advise by sector	http://tiny.cc/gdu-uk1	Comprehensive guidance documents on adapting to climate change, including the consideration of uncertainties, (in the UK and/or England) are available at these web portals
	2013	UKCIP: Tools	http://tiny.cc/gdu-uk2	
	2012	Climate Ready	http://tiny.cc/gdu-uk3	

[a]See Table 3.1 for abbreviations of countries
[b]This information is only contained in a news archive and is thus difficult to find on the web site

3.4 Conclusions

The national climate policy scene in Europe is rapidly changing. Judging by the number and breadth of national policy documents dealing with the issue, adaptation has become a mainstream activity (see also Massey and Huitema 2012). However, the perceived needs, available resources, and levels of ambition vary significantly across countries (see Table 3.1).

We can foresee a demand from the impact, vulnerability and adaptation community to deliver more sophisticated climate change scenarios. Long-term averages are no longer sufficient when more detailed questions are being asked on the nature and range of possible impacts. Short-term variability within years, the frequency and magnitude of extreme events and intermediate-term projections are gaining importance. The expanding demand for more detailed and varied climate scenarios brings uncertainties to the forefront. In this context, it needs to be emphasized that uncertainties related to non-climatic (e.g., socio-economic and technological) developments and uncertainties resulting from imperfect climate impact models are still not systematically considered in many CCIV assessments. The development of a robust knowledge base for adaptation requires increased consideration of those uncertainties, even though they cannot always be quantified.

Dealing with uncertainty is not only an academic issue but also a very practical question for planners, managers and insurance agents. Targeted guidance is needed that explains the relevance of key uncertainties and how they can be addressed by robust adaptation strategies. Organisations at the boundary between science and policy, such as the EEA, play an important role in providing policy-makers with quality-controlled information that is understandable and relevant for their specific decision context (Hanger et al. 2013). Work at the boundary between science and policy can help turning potentially useful climate information into information that is actually used by decision-makers (Lemos et al. 2012).

Dynamic interactive tools in web portals are likely to be an important part of the tool box for those who are confronted with adapting to climate change. As an example, Climate-ADAPT provides indicators on climate change, climate impacts and related vulnerabilities and a step-by-step Adaptation Support Tool. It also aims to support the learning processes between European countries by providing extensive information on the legal framework for adaptation, on the relevant knowledge base and on actual adaptation actions across Europe. If such tools can be made sufficiently user friendly, they have the advantage of supporting the mainstreaming of adaptation in various planning activities. This is important to ensure successful climate change adaptation.

We feel there is a need to develop a variety of ways of estimating and presenting uncertainties and to turn research findings into conclusions that can be used in practical applications. Addressing uncertainties in adaptation to climate change is challenging, and there is no single strategy that works best in all circumstances. Note in this context that some authors have used the metaphor of a "monster" to distinguish several strategies to cope with scientific uncertainty about climate change (van der Sluijs 2005; Curry and Webster 2011):

- "Hiding" aims at denying the existence or relevance of uncertainties;
- "Exorcism" aims at reducing or eliminating uncertainties, in particular through more research;
- "Adaptation/taming" aims at taming the monster by quantifying uncertainties;
- "Simplification" aims at standardizing the monster, e.g. by formalized IPCC guidelines for characterizing uncertainty; and
- "Assimilation" is about learning to live with the monster by rethinking one's own perspective on it, e.g. through post-normal science and other forms of reflexive science (Funtowicz and Ravetz 1992).

Each of these strategies can be recognized to some degree in the activities of the countries surveyed here. More advanced countries generally pursue several strategies in parallel, as can be shown by the example of the United Kingdom. Fundamental research sponsored by the Natural Environmental Research Council (NERC) aimed at reducing uncertainties through improved data collection and process understanding can be regarded as "monster exorcism"; the development of the probabilistic UKCP09 climate scenarios can be regarded as "taming"; classifying the confidence in specific risk projections according to three categories (low, medium and high) in the *Summary of the Key Findings from the UK Climate Change Risk Assessment 2012* can be regarded as "simplification"; and the provision of comprehensive guidance documents about living with these uncertainties (see Table 3.5) can be regarded as "assimilation".

The survey results presented here indicate that there is still plenty of work in order to convey meaningful messages on uncertainties. Dynamic interactive tools in web portals are likely to be an important part of the tool box of those who are confronted with issues related to adaptation to climate change.

References

Budescu, D.V., S. Broomell, and H.H. Por. 2009. Improving communication of uncertainty in the reports of the intergovernmental panel on climate change. *Psychological Science* 20(3): 299–308.

Curry, J.A., and P.J. Webster. 2011. Climate science and the uncertainty monster. *Bulletin of the American Meteorological Society* 92(12): 1667–1682.

Delta Programme. 2011. *Delta Programme 2012: Working on the delta*. The Hague: Ministry of Infrastructure and the Environment, and Ministry of Economic Affairs, Agriculture and Innovation. Available at: http://www.deltacommissaris.nl/english/Images/11%23094%20 Deltaprogramma%202012_EN_def%20internet_tcm310-307583.pdf.

Deltaprogramma. 2011. Deltascenario's. Verkenning van mogelijke fysieke en sociaaleconomische ontwikkelingen in de 21ste eeuw op basis van KNMI'06- en WLO-scenario's, voor gebruik in het Deltaprogramma 2011–2012. Available at: http://www.deltaportaal.nl/media/uploads/files/ Factsheet_Deltascenarios.pdf.

EC. 2013. *Guidelines on developing adaptation strategies*. Brussels: European Commission. Available at: http://ec.europa.eu/clima/policies/adaptation/what/docs/swd_2013_134_en.pdf.

EEA. 2012. *Climate change, impacts and vulnerability in Europe 2012. An indicator-based report*. Copenhagen: European Environment Agency. Available at: http://www.eea.europa.eu/publications/climate-impacts-and-vulnerability-2012.

EEA. 2013. *Adaptation in Europe*. Copenhagen: European Environment Agency. Available at: http://www.eea.europa.eu/publications/adaptation-in-europe.

Fischhoff, B. 2011. Applying the science of communication to the communication of science. *Climatic Change* 108(4): 701–705.

Funtowicz, S.O., and J.R. Ravetz. 1992. Three types of risk assessment and the emergence of post-normal science. In *Social theories of risk*, 251–273. Westport: Greenwood.

Füssel, H.-M., and R.J.T. Klein. 2006. Climate change vulnerability assessments: An evolution of conceptual thinking. *Climatic Change* 75: 301–329.

Hanger, S., et al. 2013. Knowledge and information needs of adaptation policy-makers: a European study. *Regional Environmental Change* 13(1): 91–101.

IPCC. 2005. Guidance note for lead authors of the IPCC fourth assessment report on addressing uncertainties. Available at: https://www.ipcc-wg1.unibe.ch/publications/supportingmaterial/uncertainty-guidance-note.pdf.

Lemos, M.C., C.J. Kirchhoff, and V. Ramprasad. 2012. Narrowing the climate information usability gap. *Nature Climate Change* 2(11): 789–794.

Lorenz, S., et al. 2013. The communication of physical science uncertainty in European National Adaptation Strategies. *Climatic Change* (online). Available at: http://www.see.leeds.ac.uk/fileadmin/Documents/research/sri/workingpapers/SRIPs-43.pdf, doi: 10.1007/s10584-013-0809-1.

Massey, E., and D. Huitema. 2012. The emergence of climate change adaptation as a policy field: the case of England. *Regional Environmental Change* 13(2): 341–352.

Mastrandrea, M.D., et al. 2010. Guidance note for lead authors of the IPCC fifth assessment report on consistent treatment of uncertainties. Available at: http://www.ipcc.ch/pdf/supporting-material/uncertainty-guidance-note.pdf.

Meehl, G.A., et al. 2007. Global climate projections. In *Climate change 2007: The physical science basis*, ed. IPCC, 747–845. Cambridge: Cambridge University Press.

Moser, S.C. 2010. Communicating climate change: history, challenges, process and future directions. *Wiley Interdisciplinary Reviews: Climate Change* 1(1): 31–53.

Moss, R.H., and S.H. Schneider. 2000. Uncertainties in the IPCC TAR: Recommendations to lead authors for more consistent assessment and reporting. In *Guidance papers on the cross-cutting issues of the third assessment report of the IPCC*, ed. R. Pachauri, T. Taniguchi, and K. Tanaka, 33–51. Geneva: WMO.

Nakicenovic, N., and R. Swart (eds.). 2000. *Special report on emissions scenarios*. Cambridge: Cambridge University Press.

Pfenninger, S., et al. 2010. Report on perceived policy needs and decision contexts, Mediation Project. Available at: http://mediation-project.eu/output/downloads/d1.1-2011-02-final.pdf.

Pidgeon, N., and B. Fischhoff. 2011. The role of social and decision sciences in communicating uncertain climate risks. *Nature Climate Change* 1(1): 35–41.

Rabinovich, A., and T.A. Morton. 2012. Unquestioned answers or unanswered questions: beliefs about science guide responses to uncertainty in climate change risk communication. *Risk Analysis: An Official Publication of the Society for Risk Analysis* 32(6): 992–1002.

Smith, J.B., J.M. Vogel, and J.E. Cromwell. 2009. An architecture for government action on adaptation to climate change. An editorial comment. *Climatic Change* 95(1/2): 53–61.

Steinemann, M., and J. Füssler. 2012. Workshop on experiences with climate related risk and vulnerability assessments in Europe – Documentation. Available at: http://www.bafu.admin.ch/klimaanpassung/11529/11578/index.html?lang=de&download=NHzLpZeg7t,lnp6I0NTU042l2Z6ln1acy4Zn4Z2qZpnO2Yuq2Z6gpJCHdH12f2ym162epYbg2c_JjKbNoKSn6A–.

Swart, R., et al. 2009. Agreeing to disagree: uncertainty management in assessing climate change, impacts and responses by the IPCC. *Climatic Change* 92(1–2): 1–29.

Tang, S., and S. Dessai. 2012. Usable science? The U.K. climate projections 2009 and decision support for adaptation planning. *Weather, Climate, and Society* 4(4): 300–313.

Van der Sluijs, J. 2005. Uncertainty as a monster in the science-policy interface: four coping strategies. *Water Science and Technology: A Journal of the International Association on Water Pollution Research* 52(6): 87–92.

Wilby, R.L., and S. Dessai. 2010. Robust adaptation to climate change. *Weather* 65(7): 180–185.

Chapter 4
Showcasing Practitioners' Experiences

Annemarie Groot, Ana Rovisco, and Tiago Capela Lourenço

Key Messages

Twelve real-life cases show how policy-makers, decision-makers and researchers have struggled together to deal with uncertainty in adaptation decision-making. Some key features are as follows:

- Most real-life cases conscientiously addressed uncertainties related to the use of scenarios. Few cases dealt with statistical uncertainty and /or recognized ignorance.
- In all cases a combination of multiple methods is applied to address uncertainty. In most of the cases these include expert elicitation, stakeholder involvement and sensitivity analysis.
- The cases all show that conscientiously addressing uncertainty had an effect on the adaptation decision taken and/or changed attitudes to climate change adaptation.
- Most cases show a clear shift in thinking from a deterministic or 'single optimal solution' approach to adaptation towards a flexible, robust and no-regret approach.

A. Groot (✉)
Alterra – Climate Change and Adaptive Land and Water Management,
Wageningen University and Research Centre,
Droevendaalsesteeg 3A, 6708 PB Wageningen, Gelderland, The Netherlands
e-mail: annemarie.groot@wur.nl

A. Rovisco • T. Capela Lourenço
Faculty of Sciences, CCIAM (Centre for Climate Change, Impacts, Adaptation
and Modelling), University of Lisbon, Ed. C8, Sala 8.5.14,
1749-016 Lisbon, Portugal
e-mail: acrovisco@fc.ul.pt; tcapela@fc.ul.pt

4.1 Introduction

This chapter describes real-life cases showing how policy-makers, decision-makers and researchers have struggled together to deal with uncertainty in adaptation decision-making (Fig. 4.1). We selected these case studies through a world-wide call for practical examples of adaptation decision-making processes and dealing with climate-related uncertainties. Out of the 27 real life stories, that were submitted in a prescribed format, 12 illustrative cases were selected by a group of experts, all of them members of the CIRCLE-2 Joint Initiative on Climate Uncertainties.[1] The key selection criteria were whether the story increased understanding of handling uncertainty in adaptation planning and implementation, and whether the case showed the impact of conscientiously addressing climate uncertainties on the decision taken. Other criteria for selection included: the link to a real adaptation decision-making process, the involvement of different stakeholders, and diversity in scope (geographical, sectorial and scale).

Despite some bias towards Water Management, Infrastructure and Disaster Risk Reduction projects, the cases show a wide range of decision-making processes to address climate change impacts. Only two cases show a clear single-sector focus, while all others report a multi-sector approach involving agriculture, health, biodiversity, energy and finance. All the case study initiatives are publicly funded and present a clear geographical bias towards Europe (10 cases out of 12). This is due to the fact that although we strived for an open submission of case studies and different international networks and websites were used, we mainly approached potential authors via the European network CIRCLE-2, different European research programmes, and national research programmes such as Knowledge for Climate (The Netherlands), Climate Change-Snowll (Austria) and Klimzug (Germany). Five cases describe how uncertainty is addressed at the national scale, two cases at the sub-national scale and five at the local scale (see Table 4.1). Since adaptation is a relatively new field, most of the decision-making processes deal with (policy) plans, while the actual implementation is still some years down the line. Consequently, the uncertainties dealt with in the cases are predominantly related to assessment of climate change impacts and vulnerability. Very few cases explicitly address uncertainties as to the appraisal of adaptation measures or implementation of adaptation.

The stories are constructed on the basis of interviews with the main decision-maker and the principal scientist involved, together with information on the case study provided in the submission stage. Each description highlights the challenge the decision-maker was facing, the types of climate uncertainties addressed, methods that are used to deal with uncertainties and the final decisions taken. All case studies show how the process of conscientiously addressing climate uncertainties has affected these decisions. Two types of decision making are distinguished

---

[1] This initiative is a coordinated transnational funding effort, within the scope of CIRCLE-2, aiming at sharing and advancing scientific knowledge and practice on dealing and communicating climate and climate change uncertainties in support of adaptation decision-making. More information on the Initiative is available at: http://www.circle-era.eu/np4/P_UNCERT.html

4 Showcasing Practitioners' Experiences

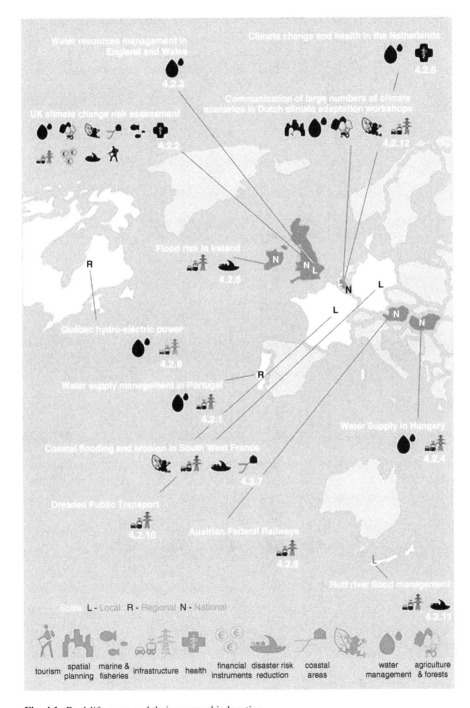

Fig. 4.1 Real-life cases and their geographic location

Table 4.1 Selected case studies according to their geographical location, sector, scale and type of decision-maker distribution

Case	Country	Sectors and Domains	Scales				Decision maker	
			National	Regional	Local	Private	Public	
Water Supply Management in Portugal (4.2.1)	Portugal	🌧️					•[a]	
UK Climate Change Risk Assessment (4.2.2)	United Kingdom	🌧️ 🍃 ☁️ 💣 ✚ 🏠 €€ 🚶	•				•	
Water Resources Management in England and Wales (4.2.3)	United Kingdom	🌧️ 🏠					•	
Water Supply in Hungary (4.2.4)	Hungary	🌧️ 🏠	•				•	
Climate Change and Health in The Netherlands (4.2.5)	Netherlands	🌧️ ✚	•				•	
Flood Risk in Ireland (4.2.6)	Ireland	🌧️ 🏠 🐟	•				•	
Coastal Flooding and Erosion in South West France (4.2.7)	France	🌧️ 🍃 🏳️ 🏠 🐟			•		•	

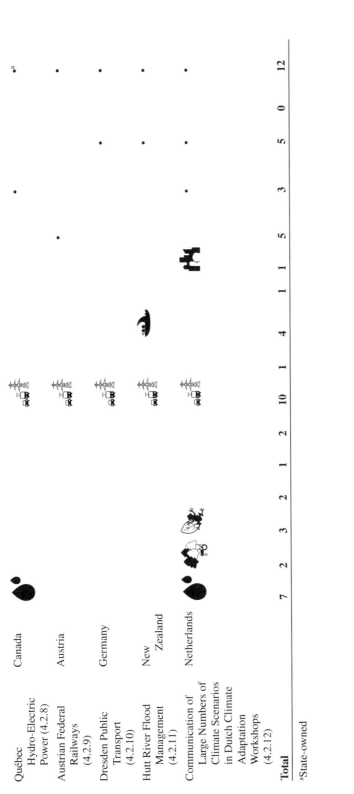

i.e. strategic and operational. Strategic decisions are fundamental and directional, and over-arching. Operational decisions, on the other hand, primarily affect the day-to-day implementation of strategic decisions. While strategic decisions usually have longer-term implications, operational decisions usually have immediate (less than 1 year) implications.

4.2 Real Life Case Studies

4.2.1 Water Supply Management in Portugal

Country: Portugal
Sector: 💧 🚆
Scale: Regional
Organisation: Public (State-owned)
Decision-type: Strategic

Key Messages

This study examined a variety of uncertainties to determine the vulnerability of a Portuguese water supply company to climate change and developed an adaptation strategy to deal with these vulnerabilities.

Key messages from the project were:

- Decision makers and stakeholders needed to be continuously involved for the success of the project. A high level of trust, generated by time-consuming engagement between the parties was necessary to deal with different views on the topic, and the company's confidential data and internal processes.
- Transferability of know-how on the topic between practitioners and researchers was critical and organisations should be able to share this knowledge.
- Quantifying cumulative uncertainty was achievable and important to support decisions, when clear criteria were agreed from the start and properly communicated.

Background

Empresa Portuguesa das Águas Livres (EPAL) is a Portuguese state-owned water utility company. It supplies about three million people living in 35 municipalities on the north bank of the Tagus River, representing more than a quarter of the Portuguese

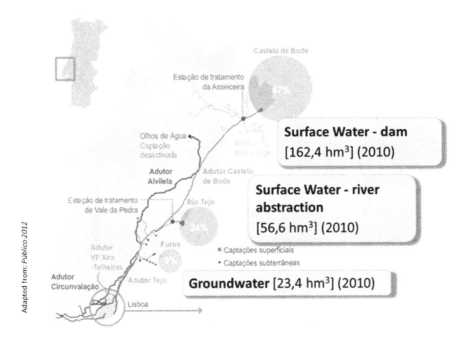

Fig. 4.2 EPAL's geographical system

population. It has three main sources of water: a large reservoir as the prime water source (67 %), the Tagus river (24 %) and groundwater from several boreholes (9 %). Further details are given in Fig. 4.2.

The purpose of the project was to: (i) assess potential climate and demand changes in the geographical area served by the water utility; (ii) identify climate change impacts on the company's water sources; (iii) assess system vulnerabilities, and (iv) identify and appraise a set of potential adaptation options and measures.

The project originated within the company's executive board, because the water sector is seen as one of those potentially most affected by climate change in Portugal. EPAL is conscious of its responsibilities to take climate change into consideration because its main aim is "to supply water, now and in the future, every day, all year round, with the necessary quality and at an acceptable cost". The project began in October 2010 and ran until May 2013.

Coordination of the project was provided by the Faculty of Sciences of the University of Lisbon and involved three other Portuguese universities. From the company's side, there was involvement from EPAL's technical and management staff (one project management committee and one advisory committee) providing company systems data and feedback on the results from the demand scenarios, impact models and other scientific information. Out of 100 of the company's key

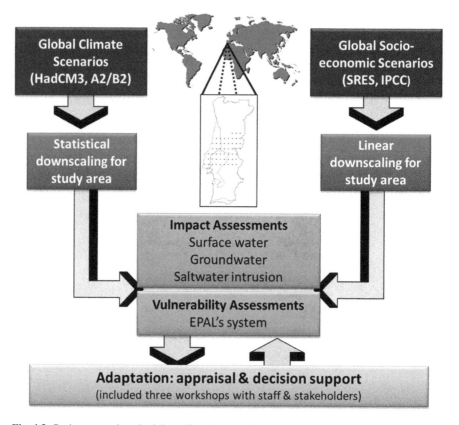

Fig. 4.3 Project general methodology. Top-down and bottom-up approach

external stakeholders (e.g. governmental, regulator, shareholders, clients, NGOs, utilities) about 20 were invited to specific meetings.

Process

The project methodology is shown in Fig. 4.3. Focussing on the development of an adaptation strategy, the project initially reviewed existing global climate and socioeconomic scenarios and downscaled these to suit the company's geographical and time scales. In the past EPAL has considered non-climatic information, such as changes in demographics, and projections on water availability have been incorporated into the project's impact assessments on surface and groundwater resources. In this study scenarios have been utilised. These include climate scenarios (e.g. precipitation) affecting water supply, and socioeconomic scenarios (e.g. demographics) affecting demand. Using these scenarios, impacts

Fig. 4.4 EPAL's Adaptation workshop

on surface water sources, groundwater sources and salt-freshwater interfaces in estuaries were modelled in terms of water quantity and quality. Vulnerability was then assessed by analysing EPAL's capacity to adapt to the potential impacts.

Climate data used

- Interpolated data from European Climate Assessment & Dataset with a grid of 25×25 km
- NCEP reanalysis data for calibration and model validation
- Coupled atmosphere-ocean general circulation model (HadCM3) downscaled using a generalised linear model
- Climate change storylines with quantitative information for socio-economic scenarios A2 and B2 (SRES) to the middle and end of century.

Three workshops were held where the results were presented, discussed and some decisions were validated. These meetings aimed to analyse the main results of the project in terms of potential impacts and adaptation measures, identifying potential synergies, conflicts and trade-offs between different alternatives and different stakeholders (Fig. 4.4).

In the last workshop, each potential impact was labelled with a level of scientific confidence (inversely correlated with uncertainty level) in order to better support the decision. To prioritise the adaptation measures for each potential impact and vulnerability, a gaming-like approach was developed. Participants were divided into smaller groups and had to choose from a set of adaptation measures (in the form of adaptation cards, previously co-created and characterised together with EPAL staff via a parallel participatory approach that focused on the adaptation objectives) and discuss the final results.

Overal, over 50 of EPAL's staff and about 20 different external stakeholders participated in the workshops. Contact is being maintained with a sample of these institutions to obtain their feedback and further understand their influence on EPAL's adaptation processes. The majority of the adaptation measures, for example the reduction of pollution in aquifers, need the support of external stakeholders, and the feasibility of measures is being discussed with them.

Continuous interaction with the two internal project committees was designed, among other objectives, to help EPAL's staff and stakeholders understand the meaning of uncertainty in the context of climate adaptation decision-making.

Uncertainty Assessment

Within each project phase different levels of uncertainty were acknowledged and considered for each of the project's activities:

Example of handling uncertainty in hydrological impact modelling using a sensitivity analysis

EPAL is concerned that the freshwater-saltwater interface along the Tagus River estuary could reach its abstraction point at Valada (about 32 km upstream) through a potential combination of reduced river discharge, sea level rise and salinity increases. This would either require the implementation of adaptation measures such as nanofiltration, or the abandonment of the facilities. Past assessments place the interface 15 km downstream of EPAL's abstraction point and a numerical simulation model (CE-Qual-W2) was used to evaluate the potential impacts. However, consultation with the company's experts revealed that the complexity around the river-estuary-sea system created extra uncertainty and reduced their confidence in the model results. A sensitivity analysis using additional model runs was undertaken and results supported, with a high level of confidence, that significant salt water intrusion is not to be expected. Thus, the companies' decision was to not advance with specific adaptation measures at this time.

- Selection of scenarios,
- Socioeconomic data downscaling,
- Climate data downscaling,
- Hydrological and hydrogeological impact modelling,
- Vulnerability assessments,
- Adaptation options appraisal.

High Agreement	High Agreement	High Agreement
Limited Evidence	Medium Evidence	Robust Evidence
Medium Agreement	Medium Agreement	Medium Agreement
Limited Evidence	Medium Evidence	Robust Evidence
Low Agreement	Low Agreement	Low Agreement
Limited Evidence	Medium Evidence	Robust Evidence

Agreement ↑ — Evidence →

Confidence Scale: Robust, Limited

Figure: Confidence integrates models and scenarios Agreement and current and future Evidence. Confidence increases towards the top-right corner as suggested by the increasing strength of shading and it was expressed in three qualifiers: Limited, Medium and Robust (Adapted from "Guidance Note for Lead Authors of the IPCC Fifth Assessment Report on Consistent Treatment of Uncertainties", 2010)

Fig. 4.5 Confidence levels used to communicate uncertainties to decision-makers

Based on current adaptation literature, uncertainties within these activities were dealt with in the following ways:

- Scenario analysis,
- Expert elicitation,
- Sensitivity analysis,
- Stakeholder involvement,
- Extended peer review (review by stakeholders).

From the beginning, the various scientific teams were asked to qualify the uncertainties in their results. Each potential impact was then communicated and associated with a level of confidence derived from a balance between the level of agreement (with other comparative studies) and the level of evidence (statistic robustness of models; quality of observed data) (Fig. 4.5).

The uncertainties associated with the impact of competition between EPAL and other organisations on water resources were not taken into account in a quantitative way (i.e. via models), but addressed through the involvement of stakeholders and expert elicitation of 'what if' issues.

Effect of Uncertainty on Decision–Making

From the start of the project it was clear that not all of EPAL's staff involved had the same attitude to the climate change topic and level of confidence on the potential results of the vulnerability assessment. This is partly because they come from different areas within the company and so have different perspectives regarding the role of risk and uncertainty in operational and strategic decisions. In practical terms this meant that some EPAL staff members felt that for some decisions, despite uncertainties, there was enough confidence in the results, while for other results there was a need to further reduce those uncertainties. For other EPAL staff members still, uncertainty was deemed to be too large for results to provide sufficient support to decisions.

For example, quantity and quality water issues in the *Castelo de Bode* dam (primary source of water to the system) due to changes in temperature, precipitation and stream flow were modelled using two sets of emissions scenarios (A2 and B2). This provided information to support decisions on the strategic use of the reservoir relative to other available sources in the future. It also inspired the creation of a protocol with EDP (a large electricity company that utilises the same water source) to agree on rules for the use of water in years of scarcity. However, the reservoir is located in an area prone to forest fires that may require adaptation efforts to prevent such wildfires. Despite the efforts of researchers it was not possible to model the physical interactions of such fire events and their consequences on water quality. Significant uncertainties still remain and no decisions on specific adaptation options were made. This contrasted with the work carried out for the Valada abstraction point (see box on 'dealing with uncertainty') that accounts for about one quarter of EPAL's supplied water. In this case the confidence of EPAL's decision-makers was improved through further analysis to enable them to make decisions on investments in the Water Treatment Plant associated with the abstraction point, such as not to install a nano-filtration system in the near future.

Finally, an adaptation strategy has been prepared, including a diagnosis of EPAL's current and future climate related vulnerabilities, and a set of priotized adaptation options. The strategy was designed to accommodate a general no-regret approach but for some decisions the precautionary principle was applied. The strategy is designed to support decisions on which adaptation options or sequences of adaptation measures (pathways) are better able to cope with the current and future vulnerability. The chosen options are expected to be mainstreamed into EPAL's regular management and strategic planning and can also serve the company in its relationship with external stakeholders. The strategy's implementation is to be monitored by the company and revised every 5 years.

Authors: David Avelar, Tiago Capela Lourenço and Ana Luis

Links for more information: http://siam.fc.ul.pt/adaptaclima-epal/?lang=en, www.epal.pt

Contact details: dnavelar@fc.ul.pt, tel.: +351 217 500 939

4.2.2 UK Climate Change Risk Assessment

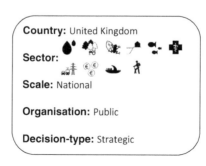

4 Showcasing Practitioners' Experiences

Key Messages

The UK Climate Change Risk Assessment (CCRA) was the first-ever comprehensive assessment of the potential risks and opportunities arising as a result of climate change in the UK. The results of the Climate Change Risk Assessment are being used by a variety of government departments in Scotland, Wales and Northern Ireland to facilitate comparisons across sectors, prioritise adaptation actions and improve confidence in decision-making.

Key messages from the CCRA were:

- Despite uncertainties, evidence is now sufficient to identify a range of possible climate change impacts and indicate their relative magnitude to inform adaptation and planning.
- Decision-making needs to consider uncertainties in order to identify robust options.
- Presenting the full spread of results to stakeholders through the use of the "score cards" was a useful way of communicating uncertainty.
- Flexibility needs to be built into adaptation planning to allow for a future climate that may change more slowly, more quickly or in a different way than currently expected.
- The use of "sector champions" appeared to be a useful approach to involve relevant stakeholders in the assessment of risks, including the management of related uncertainties.
- Climate change is only one driver amongst many and should be considered alongside other drivers when assessing future risk.

Background

The UK Climate Change Act 2008 made the UK the first country in the world to have a legally binding, long-term framework to cut carbon emissions and develop adaptation strategies?. As a response to this, the UK government set up the first CCRA, which was reported in 2012 and is scheduled to be updated every 5 years to take into account new data and improved understanding of the issues. This first report outlined some of the most important risks and opportunities presented by climate change across 11 sectors. By analysing existing data, impacts were assessed for three time slices and across three emission scenarios.

The consortium[2] carrying out the review was supported by leading technical experts in the 11 sectors who acted as "sector champions". The aim was to build a consistent picture of risk across the UK and allow for some comparison between disparate risks

---

[2] HR Wallingford led a consortium consisting of the Met Office, AMEC Environment & Infrastructure UK Ltd, Collingwood Environmental Planning, Alexander Ballard Ltd, Paul Watkiss Associates and Metroeconomica, in order to carry out the review. Sector champions included Cranfield University, CEFAS, Forestry Research, Birmingham University, Acclimatise, the Hutton Institute and the Centre for Ecology and Hydrology.

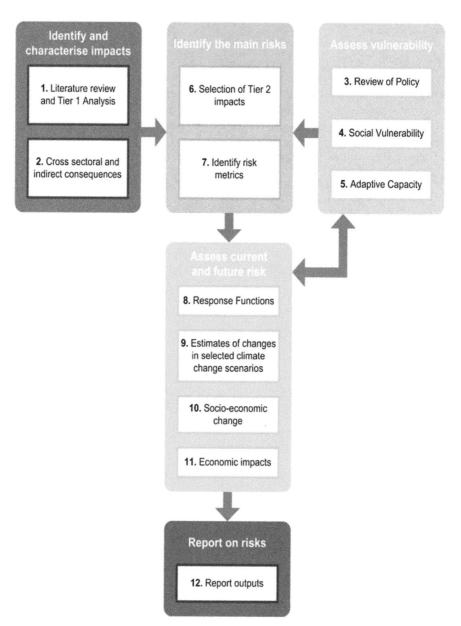

Fig. 4.6 Steps involved in producing the CCRA

and regional/national differences. The UK government was the primary 'customer' for the CCRA although the assessment engaged more than 1,800 stakeholders through workshops, online questionnaires and report reviews. These stakeholders came from a wide variety of backgrounds, including non-governmental organisations, leading businesses within sectors, regulatory bodies and government agencies and were involved in identifying and prioritising risks. They also reviewed draft outputs to ensure that the information presented was both understandable and useful.

Process

The steps involved in producing the assessment are described in Fig. 4.6.

Over 700 impacts of climate change were identified (Tier 1) across the 11 sectors under review. These were combined with an assessment of vulnerability across the UK as a whole to identify the main risks. As part of this, a 2nd tier of about 100 impacts was extracted using a simple multi-criteria scoring system based on the magnitude of consequences, likelihood of occurrence and urgency of decision required.

For each impact in the Tier 2 list, one or more risk metric(s) was identified. These provided measures of the consequences of climate change, relative to specific climate variables.

Examples of risk metrics linked to the impact "major drought"

- Reduced summer river flow
- Change in public water supply availability
- Population in areas with future water supply deficits

The next step was to develop response functions, being the relationship between a risk metric (e.g. crop yield) and one or more climate variables (e.g. temperature or precipitation). Response functions were derived in a number of ways:

- Sensitivity analysis of detailed models,
- Historical data to produce a simple statistical relationship,
- Expert elicitation where models or data was not available.

Climate data used

- UKCP09/UKCIP02 projections
- Met Office observed weather and climate
- Hadley Centre HadCM3 (sea ice)
- Low to high emission scenarios
- UKCP09 probability levels

Uncertainties associated with these approaches were taken into consideration as part of the overall confidence scoring for each risk metric. The magnitude of climate risks were then analysed using climate projections for three time slices and three emissions scenarios:

- 2020s (2010–2039) – medium emissions scenario,
- 2050s (2040–2069) – low, medium and high emissions scenario,
- 2080s (2070–2099) – low, medium and high emissions scenario.

It was recognised that many of the risk metrics in the CCRA were influenced by a wide range of drivers other than climate change. For example, risks related to flooding, water supply and demand, health and energy demand were particularly sensitive to future population and a standard set of population projections were applied to across all sectors.

Uncertainty Assessment

Uncertainties were considered in the following areas:

- **Climate system**: driven by limitations in our ability to model certain aspects of the climate system, as well as intrinsic modelling uncertainty and the nature of the system.
- **Future emissions**: captured within the UKCP09 projections that were used in the CCRA to project the risk moving into the future.
- **Current level of risk faced**: particularly important in relation to extreme events, the estimation of which was also subject to considerable uncertainty.
- **The relationship of the risks to climate variables**: through models, statistical relationships and the use of simple 'response function' relationships.
- **Planned or autonomous adaptation and changes in society** (social and economic): assumptions were made on a case by case basis. Population projections were applied but the vast majority of the work in the CCRA took this as a qualitative consideration.
- **Financial consequences** of impacts could only be estimated as part of a monetisation exercise, for example the intrinsic value of elements of the natural environment was not captured.

These uncertainties were handled, amongst others, in the following ways:

- **Emission scenario analysis.** Within each projection a probabilistic range was used, from the 10th percentile to the 90th percentile probability level. Population projections (low, principle and high) were also applied to provide results combining both climate and population changes.
- **Expert elicitation** and **peer-review** were utilised to substantiate whether the assumptions adopted were reasonable.
- **Stakeholder involvement** was utilised to ensure that uncertainties presented in reports were understandable to the reader.

One key method of presenting results to stakeholders, to generate an appreciation of uncertainty, was through the use of "score cards". The risk metrics considered in this first CCRA varied in character and whilst some were quantified, others had to rely on expert elicitation, or a narrative based on the literature. To allow comparison of these different risks, they were categorised as having either 'high', 'medium' or 'low' magnitude consequences and either a 'high', 'medium' or 'low' confidence. An example for agriculture and forestry is shown in Fig. 4.7. This shows the lower (l), central (c) and upper (u) estimates of magnitude of the consequences (based on the range of emissions scenarios analysed and associated probability levels) for the three time slices considered (i.e. the 2020s, 2050s and 2080s) and the overall level of confidence in these estimates (L – Low, M – Medium or H – High).

Metric code	Potential risks for agriculture and forestry	Confidence	Summary Class 2020s			2050s			2080s		
			l	c	u	l	c	u	l	c	u
AG1b	Changes in wheat yield (due to warmer conditions)	M	1	2	2	2	2		2		
AG9	Opportunities to grow new crops	H	1	1	1	2	2	2		1	
AG1a	Changes in sugar beet yield (due to warmer conditions)	M	1	1	2	1	2		2		
AG10	Changes in grassland productivity	M	1	1	1	1	2	2	1	2	2
FO4b	Increase of potential yield of Sitka spruce in Scotland	M	1	1	1	1	1	1		1	
AG1c	Changes in potato yield (due to combined climate effects and CO₂)	L	1	1	2	1	1	2	1	1	2
FO1a	Forest extent affected by red band needle blight	M	1	2			2			2	
AG11	Increased soil erosion due to heavy rainfall	L	1	2	2	1	2		1		
AG5	Increases in water demand for irrigation of crops	M	1	2		1	2		2	2	
AG4	Drier soils (due to warmer and drier summer conditions)	M	1	2	2	1	2		1	2	
AG2a	Flood risk to high quality agricultural land	H	1	1	2	1	2	2	2		
FO4a	Decline in potential yield of beech trees in England	M	1	1	1	2	2	2			
BD12	Wildfires due to warmer and drier conditions	M	1	1	2	1	2		2	2	
FL14a	Agricultural land lost due to coastal erosion	H	1	1	1	1	2	2	2	2	
WA8a	Number of unsustainable water abstractions (agriculture)	M	1	1	2	1	2	2	2	2	2
FO1b	Forest extent affected by green spruce aphid	M	1	1	2	1	2	2	1	2	
FO2	Loss of forest productivity due to drought	M	1	1	2	1	1	2	1	2	
AG8b	Dairy livestock deaths due to heat stress	L	1	1	2	1	1	2	1	1	2
AG7b	Reduction in dairy herd fertility due to heat stress	L	1	1	2	1	1	2	1	1	2
AG8a	Increased duration of heat stress in dairy cows	H	1	1	1	1	1	2	1	1	2
AG7a	Reduction in milk production due to heat stress	L	1	1	1	1	1	1	1	1	
AG3	Risk of crop pests and diseases	L	Too uncertain								

M	Confidence assessment from low to high
	High consequences (positive)
2	Medium consequences (positive)
1	Low consequences (positive)
1	Low consequences (negative)
	Medium consequences (negative)
	High consequences (negative)
~	No data

Fig. 4.7 Score card indicating the consequences and confidence levels of risk metrics under climate change in the agricultural and forestry sector

For example, metric AG1b "Changes in wheat yield (due to warmer conditions)" is projected (with medium confidence) to have low to medium positive consequences by the 2020s and medium to high positive consequences by the 2050s and 2080s. This can be compared with metric AG10 "Changes in grassland productivity", where it is projected (with medium confidence) to have low positive consequences by the 2020s and low to medium positive consequences by the 2050s and 2080s. Therefore, the score card shows not only shows the scale of the consequences (i.e. low, medium or high), but also the range in uncertainty of the projections (from l – lower, to c – central and u – upper projections) and the speed of onset of consequences (i.e. by the 2020s, 2050s or 2080s). It has been deliberately chosen to use the same colour for both the low positive and low negative consequences. The score card helps the decision-makers to prioritise areas of action by comparing the relative magnitude of risks and indicating how soon action should be taken to mitigate or adapt to that risk.

Fig. 4.8 The M1 and River Trent valley on 10 November 2000 (Source: Frameworks for delivering regular assessments of the risks and opportunities from climate change: An independent review of the first UK Climate Change Risk Assessment. Final Report, 18 June 2012 Robert L. Wilby)

Effect of Uncertainty on Decision–Making

"There is a risk of being locked into maladaptation"

The reports produced from the CCRA reflected potential risks and opportunities and did not purport to be a prediction of the future consequences of climate change. Despite uncertainties over the magnitude and timing of climate change impacts, the CCRA was able to provide sufficient evidence to identify a range of possible outcomes that can inform adaptation policies and planning.

The results are being used by UK government departments and devolved governments as part of their evidence base to support decision-making on adaptation to climate change in organisations across the country. Decision-makers recognise that they need to consider uncertainties and to allow flexibility in their policies and plans, and they need to report their actions under the "Adaptation Reporting Power" of the Climate Change Act 2008. Decisions range from the simple "low cost, no regret" measures, such as urban greening, through to the adaptation pathway approach, in which flexibility is maintained and adjustments made if conditions or information change. An example of the latter is the Thames Estuary 2100 project being a multi-million pound contract planning for flood risks in London. The CCRA provides a probabilistic climate change framework with differing degrees of confidence over various outcomes to facilitate this decision-making process (Fig. 4.8).

Author: Helen Udale-Clarke

Links for more information: http://www.defra.gov.uk/environment/climate/government/risk-assessment/

Contact details: h.udale-clarke@hrwallingford.com, tel: 01491 822325

4.2.3 Water Resources Management in England and Wales

Country: United Kingdom
Sector: 💧
Scale: Local
Organisation: Public
Decision-type: Strategic

Key Messages

"The effects of climate change uncertainties are not as immediate as issues such as changing water demand"

This project stemmed from the desire of the Environment Agency of England and Wales to account for the large uncertainties in climate change projections in planning water requirements of the future.

Key messages from this work were:

- Planning based on just a few storylines was a risk in itself.
- There was a need for water management options that are flexible and robust under a range of possible futures.
- Tools, such as large climate model ensembles in combination with risk based decision-making frameworks, can be used to avoid poor adaptation decisions.

Background

This research project was commissioned by the Environment Agency of England and Wales and initially carried out by the School of Geography and the Environment, Oxford University. Every 5 years, water companies have to indicate how they will guarantee the supply of water over the following 25 years. The Environment Agency wanted to provide guidelines to water companies on how to take into account large uncertainties in climate change information when preparing the associated 5 year Water Resources plans.

Water companies in England and Wales have considered the impact of climate change in their plans since 1998, but approaches tend to be simple and deterministic, as climate change is one of many factors that companies have to take into account. The Environment Agency wanted to explore how large ensembles of climate information could be used to improve decision-making.

Apart from the Environment Agency, other stakeholders included managers from some of the water companies, climate scientists, and hydrologists. All of these were consulted during the development of the project.

Process

"Tools need to be simple and cheap"

The project concentrated on exploring climate model related uncertainties as represented by the climate data described on the box.

Climate data used

- Perturbed physics ensemble (PPE) – 247 members – based on the HADCM3 model
- An ensemble of opportunity consisting of 21 General Circulation Models (GCMs) available through the CMIP3 database (IPCC 4th Assessment Report)

Both ensembles were run under the SRES A1B emissions scenario.

It was the first project to use such a large range of climate models to study the effects of climate projection uncertainties on the management of a water resources system. The Environment Agency was involved in the design of the project, the selection of hydrological modelling tools and calibration of models, and the choice of adaptation options. Workshops were also organised so that the scientists could understand the information needs of decision-makers in this sector, and determine the sort of information that could be provided.

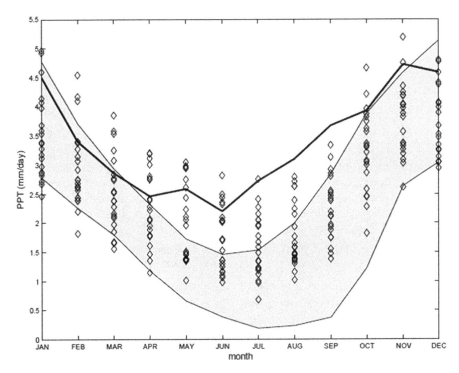

Fig. 4.9 Mean monthly precipitation (mm/day) for 1930–1984. The thick line corresponds to observed monthly means, the grey shadow indicates the range of precipitation simulated by the PPE, and the diamonds indicate the CMIP3 models results

The large ensemble of climate projections was run through a hydrological model and then a water resources model for a catchment in the South West of England, to evaluate the time dependent risk of failure to supply water demand under different adaptation options. The hydrological and water resource models were already in use by water supply companies and regulators. Since time and expense was not required to develop these tools it was hoped that they would encourage the take up of information from large ensembles of climate models.

An example of the exploration of uncertainties in climate projections can be seen in Fig. 4.9 which shows the mean monthly precipitation (mm/day) for the period 1930–1984. The fact that, in this case, uncertainties in the ranges of model physics (PPE) and model structure (CMIP3 models) do not coincide, shows that both ensembles are necessary to better explore the full range of climate model uncertainty.

Uncertainty Assessment

The primary uncertainties analysed by running the large ensemble of climate models through the water resources system model were those due to:

- climate model structure represented by the CMIP3 models,[3]
- climate model physics represented by the perturbed physics ensemble (PPE).[4]

Other sources of uncertainty such as emission scenario and impact model uncertainty were ignored in this study. It is expected that the uncertainty range might vary when all sources are taken into account.

Within the Environment Agency there was already an awareness of uncertainties in climate change risks. They became particularly interested, however, in the fact that the range of uncertainties explored by the PPE was in general larger than that expected from the CMIP3 ensemble.

Water companies find large ensembles of climate information difficult to use. As a result of this and other projects, guidance was developed in two areas:

- Translation of climate ensembles into a range of river flows being a format that is familiar to water companies. This effectively gave them a set of impact data to use.
- Guidance on how to use the data. This gives them the confidence that using the approach will result in robust decisions.

Water company representatives argued that even though they found the results interesting, they did not have the resources to implement such analysis. They also commented that climate change risks represent only a small part of the total risks they have to face. For instance, in many parts of the UK, the main problem is changes in demand due to population increase. Even though plans have to be made for 25 years into the future, climate change and climate risks may not be the most significant risk drivers. Consequently, water companies preferred the simplified idea of using a maximum of three climate scenarios (low, medium, high) to explore climate change impacts.

---

[3] http://www-pcmdi.llnl.gov/ipcc/about_ipcc.php
[4] http://climateprediction.net/

4 Showcasing Practitioners' Experiences

Example of handling uncertainty: failure of water supply

This represents the case of a water company required to meet water demand in its catchment region into the twenty-first century at a minimum cost. The top panel of Fig. 4.10 shows a histogram of the percentage change in summer average precipitation of 2050–2079 compared to 1960–1989, for the PPE ensemble.

The bottom panel of Fig. 4.10 shows, for each range of precipitation change on the top panel, the corresponding average number of failures to supply the required demand for the business as usual (BAU) scenario and four different adaptation options. The adaptation options available include increase supply (green and purple lines in bottom panel) and/or reduce demand (red and light blue lines in bottom panel). The blue line represents business as usual. Robust adaptation options are those that, for an acceptable level of risk, reduce the risk of failure across a range of plausible climates. If for instance only five failures are acceptable, only red, light blue and purple adaptation options are robust across the range of plausible futures.

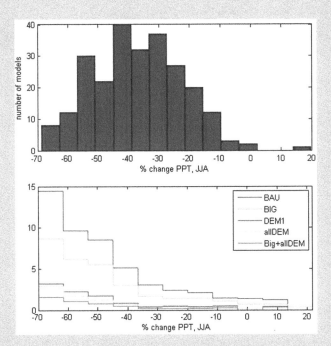

Fig. 4.10 Histogram of the percentage change in summer average precipitation of 2050–2079 compared to 1960–1989, for the PPE ensemble (*top panel*); Average number of failures to supply the required demand for the business as usual (BAU) scenario and four different adaptation options

Effect of Uncertainty on Decision–Making

"Planning based on just a few storylines is a risk in itself"

This exercise showed that using information from a small number of projections could be misleading, either over or underestimating the changes in climate risks. The Environment Agency and water companies accept that planning based on just a few storylines is a risk in itself.

From the water companies' perspective, there are many existing uncertainties other than climate change which tends to be a long-term issue. Uncertainties due to demand and environmental standards for example are much more relevant on a short-term basis. However, they appreciate the need for the use of many models and are willing to utilise the results as long as it is relatively simple to do so.

From the results produced, the Environment Agency has developed guidance on the use of probabilistic climate change information to explore sensitivity and minimise surprises for the next round of water resources plans. This will be used for the plans due to be drawn up in 2014. It will be interesting to see whether the attitude of the water companies changes after this round of plans.

Authors: Ana Lopez and Glenn Watts

Links for more information: Information about the Environment Agency guidelines for managing drought and the balance between water supply and demand can be found at http://www.environment-agency.gov.uk/business/sectors/32399.aspx

Contact details: ana.lopez@univ.ox.ac.uk, a.lopez@lse.ac.uk, tel: 44(0)7791 692025

4.2.4 Water Supply in Hungary

Country: Hungary
Sector:
Scale: Regional
Organisation: Public
Decision-type: Operational+Strategic

Key Messages

This project investigated the effects of climate change on drinking water supply in two regions of Hungary in order to support decisions on adaptation.

Key messages from the project were:

- Despite uncertainty in long-term trends of precipitation and the hydrological consequences, decisions were found to be possible.

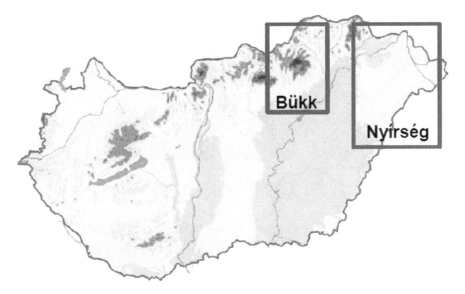

Fig. 4.11 Hungarian test areas

- As a preparation for adaptation planning, all current and future hazards should be estimated and ranked according to likelihood and severity of consequences as in the Water Safety Plan of the World Health Organization.

Background

The Hungarian National Institute for Environment (NeKI) is responsible for the water management policy of Hungary and acted as partner in the Climate Change and Impacts on Water Supply (CC-WaterS) project. The aim was to assess the climate change impacts on the future availability and safety of public water supply. In order to provide information to water managers, it considered the economic losses or benefits related to changes in climate and land use. The project was funded under the South East Europe Transnational Cooperation Programme, comprising 18 partners and was completed in May 2012.

Two specific areas located in the north-eastern part of the Hungary were analysed: the mountainous Bükk region, and the plain area of Nyírség (see Fig. 4.11). The Bükk-Mountain region encompasses the highest karstic plateau of Hungary, situated in the Carpathian Mountains. From the group of karstic springs in its South Eastern section, one large city and three villages (about 190,000 people) are supplied by one water company. The lowland area of Nyírség is part of the Great Hungarian Plains and located near the Tisza River. The mean elevation of this region ranges between 150 and 200 m. and about 260,000 people live here, settled in one large city and 60 smaller settlements. The drinking water is obtained from shallow and deep porous aquifers of the alluvial deposit and supplied by one large regional water company (84 % of the

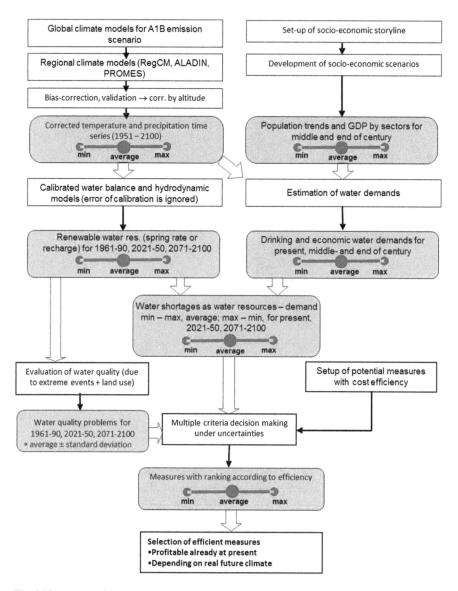

Fig. 4.12 Process of the assessment, including uncertainties (in *red colour*)

population) and a number of small waterworks. The two large water companies, representative of each region, were involved from an early stage in the study.

The main phases of the project, including the uncertainties involved are summarised in Fig. 4.12. This shows the relationships between different steps such as the establishment of climate datasets, the determination of water resources availability, estimation of water demands, evaluation of problems and selection of efficient measures, and the consideration of uncertainties (in red colour).

4 Showcasing Practitioners' Experiences

The project utilised three regional climate models (RCMs) and the SRES A1B emission scenario, with appropriate corrections (see box). To project the impact of climate change on drinking water availability and quality, the precipitation and temperature time series from the RCMs were used as input for a water balance model, a hydrodynamical model and a crop model. These models also took land use changes due to climate change into account.

Climate data sources

- SRES A1B emission scenario and three RCMs (ALADIN; RegCM and PROMES) were selected for modelling time series of temperature, precipitation and CO_2 concentration up to 2100
- The time series were bias corrected for the two pilot areas using temperature and precipitation data of E-OBS database (1961–90 period).
- Climate data was validated using observations other than those in E-OBS database. In the Bükk region correction according altitude was necessary.

Without a particular link to possible climatic futures, local experts were asked to develop a storyline showing their perceptions of the future for all social and economic aspects such as: market policy, declining and growing sectors, technical development, unemployment, governance structure, role of policy, demography, sustainability and equity. Project managers then used the storyline to develop three scenarios indicating a maximum, minimum and plausible future water demand. Experts and the two water companies were asked to provide feedback on the scenarios.

The changes in the drinking water demand were estimated on the basis of the three socio-economic and regional climate scenarios (maximum, minimum and plausible).

In the last project phase, cost-efficient adaptation measures were selected.

Uncertainty Assessment

All the stakeholders recognised uncertainties, and none of them considered them to be barriers to adaptation. Experience of very heavy precipitation in Bükk (in 2006, 2009 and 2010) and drought in both regions (beginning of 90s, 2000, 2003, and 2007) had convinced them that climate change is an issue which needs to be considered. Water management companies are not worried *whether* climate change will occur but *what* are the possible scenarios and the corresponding efficient measures.

Uncertainties of the following applied models and methods were dealt with:

- Regional Climate Models,
- Hydrological/ hydrodynamical impact models,

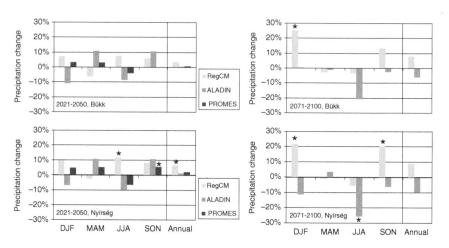

Fig. 4.13 Projected changes in seasonal mean precipitation with the use of three regional climate models, for 2021–2050 and 2071–2100. Significant changes (at 0.05 level) are indicated by asterisks (CC-WaterS, 2010 http://www.neki.gov.hu/uploads/458/Attachments/cc_waters_wp3.pdf)

- Empirical methods for estimating water balance elements (e.g. evapotranspiration),
- Land use change evaluation methods,
- Crop models for evaluation of nutrient balance elements and yields,
- Evaluation methods for socio-economic changes.

A combination of the following methods was used to address uncertainties:

- Expert elicitation,
- Sensitivity analysis of parameters, comparative analysis of formulas,
- Probabilistic multi model ensemble,
- Fuzzy Multiple Criteria Decision Making (see box).

When the project began it was expected that the results given by the three RCMs would be more or less similar but the models presented different climate changes. As can be seen in Fig. 4.13, simulations of the RCMs often do not agree even on whether the projected changes in precipitation is positive or negative. Uncertainty related to predicted seasonal precipitation with different RCMs is larger than the changes compared to the baseline. The uncertainty was more pronounced in precipitation than temperature (not shown), which shows clear and continuous increase in all seasons.

In addition, short heavy rain (causing quality problems in Bükk recently) could not be modelled which poses difficulties in planning adaptation measures against flash flood events.

Evapotranspiration seems to be the most uncertain water balance element since the parameters of the empirical formulas are perhaps not valid under considerably higher temperature. The most realistic formula was selected based on comparative analysis.

In order to draw conclusions on water availability, it was important to determine the uncertainty of climate data in water balance and hydrodynamical modelling, carrying out several simulations with various climate data. As a result, the uncertainty of the available water resources was presented as a range of possible values alongside the average values. It was noted that uncertainties in the parameters of the water balance model and hydrodynamic model were reduced through a detailed calibration procedure.

To analyse future water demand, population birth/death ratio and migration rates were projected, given envisaged economic conditions, social measures, employment and income. The impact of climate change was also considered on the likely increase in water demand for hygienic use and for watering gardens, in proportion to the increase of temperature. In this way, uncertainty in the meteorological prognosis was also incorporated in the estimation of water demand.

In the last step of the process, a Fuzzy Multiple Criteria Decision Making tool was applied to help the water companies take decisions. The best adaptation option can be selected when multiple alternatives exist even under uncertainty, represented by so called fuzzy numbers (see box).

Handling uncertainty – Fuzzy Multiple Criteria Decision Making

Fuzzy sets (representing the minimum, maximum and average values of a parameter) were used to estimate ranking criteria values e.g. cost, acceptance, flexibility and lag time and then to evaluate the composite indicator numbers. Fuzzy Decimaker version 2.0 was used as a Fuzzy Multiple Criteria Decision Making tool that helps the user to select the best solution considering a number of conflicting criteria under uncertainties.

Effect of Uncertainty on Decision-Making

Despite the fact that each of the three regional climate models gave different results, water management companies were prepared to accept the uncertainty and act. They proposed that different adaptation measures should be developed for the future range of scenarios (maximum, minimum and average). Several alternative management measures were formulated: water supply management, water demand management, shortage consequence management, change of allocation of available supply among users, water quality management and combinations of the alternatives. In the mountainous area the water management company has established a new system to monitor heavy rains and flash floods. It also intends to install a new treatment plant which can be used to protect water quality during flash floods. A proper monitoring system to measure climate and hydrological parameters was considered essential for dealing with uncertainty.

In the low lying area the regional company has begun to shut down very small water works and is trying to concentrate on larger water sources, developing a regional pipeline system in order to increase the safety of water quality. They have also made a study of prospective refuges into which they can move their operations which would make the water system less vulnerable to extreme events.

Author: Agnes Tahy and Zoltan Simonffy

Links to more information: http://www.ccwaters.eu, http://www.neki.gov.hu/?TeruletKod=0&Tipus=content&ProgramElemID=66&ItemID=458

Contact details: agnes.tahy@neki.gov.hu and simonffy@vkkt.bme.hu

4.2.5 Climate Change *and Health in The Netherlands*

> **Country:** Netherlands
> **Sector:** 💧 ✚
> **Scale:** National
> **Organisation:** Public
> **Decision-type:** Strategic

Key Messages

This case study assessed the degree of uncertainty in various potential health effects of climate change in the Netherlands.
Key lessons learned were that:

- Potential health effects due to climate change were associated with large uncertainties and knowledge gaps.
- Analysing and characterising uncertainty by means of a typology combining a scale of 'Level of precision' with 'Relevance for policy' was very useful for the selection and prioritisation of robust adaptation policies.
- Recognition of uncertainty of various health effects due to climate change had implications for policy. For example, adaptation policies that focus on enhancing the health system's capability of dealing with uncertainties were most appropriate for climate related health impacts characterised by recognised ignorance.

Background

Climate change can influence public health in many, often subtle and complex ways. Some of these potential impacts are direct, such as the impact of heat waves on heat-related deaths. Others are more indirect, such as the effect of changing climates on the distribution of vectors such as specific types of mosquitoes, which affect the

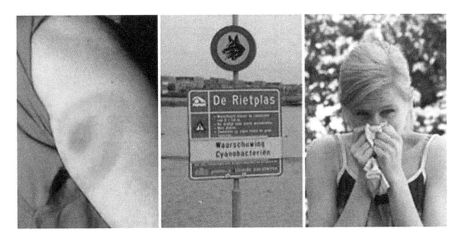

Fig. 4.14 A warning of cyanobacteria for swimmers

Fig. 4.15 The oak processionary caterpillar which entered the south of the Netherlands in the 1990s and gradually spread north. A further spread and increase in population size is expected due to climate change

distribution and risk of disease outbreaks (Figs. 4.14 and 4.15). There is a colourful mix of information on the topic, ranging from qualitative discussions on plausible impacts, through lists of knowledge gaps and research needs, to detailed quantitative studies. Projections of health risks of climate change are surrounded by uncertainties, leading to difficulties in determining the policy approach.

The Netherlands Environmental Assessment Agency (PBL), being the Dutch national institute for strategic policy analysis in the field of the environment, nature and spatial planning, has recently produced the assessments "Impacts of climate change in the Netherlands: 2012" (2012) and "Roadmap to a climate-proof Netherlands" (2009) for the Dutch government. Within these assessments it was important to account for uncertainties in a policy-relevant way and so PBL asked Utrecht University to characterise the uncertainties associated with various health effects, and to provide strategic options on how to deal with them in adaptation policy.

Process

The process carried out by the Utrecht University was as follows:

- A list of 33 potential health impacts of climate change was compiled based on existing Dutch impact assessments and international literature. These impacts were grouped into eight health themes: temperature, allergies, pests, vector-borne diseases, food/water-borne diseases, air quality, flooding/storm and UV effects.
- A questionnaire based on expert elicitation was completed. National and international experts (scientists and practitioners) were asked to indicate the level of precision with which health risks could be estimated given the present state of knowledge.
- Suggestions were made for dealing with uncertainties in climate change adaptation policy strategies.

Categories of health impacts of climate change included

- Temperature
- Allergies
- Pests
- Vector-borne diseases
- Food- and waterborne diseases
- Air quality
- Flooding and storm
- UV-related

The results of the study were used as input to PBL's impact and adaptation assessment. They were also presented at a World Health Organization (WHO) workshop on policy options for climate change and health.

Uncertainty Assessment

In the first part of the study the participating experts were asked questions to assess the 'Level of Precision' with which health risk estimates could be made given the current state of knowledge. They were also asked to provide full backup for their scores. For example:

- Why is it possible to indicate the direction of change, but not provide a quantitative risk estimate?
- What factors prevent a more precise analysis (e.g. whether data is unavailable, or cause-effect relationships not understood)?
- What factors are available that allows a certain level of precision to be applied (e.g. whether well-established models or detailed data sets are available)?

Example of handling uncertainty: 'Level of Precision' scale

The following 'Level of Precision' scale was used to assess the degree to which health effects of climate change can be quantified:

1. Effective ignorance
2. Ambiguous sign or trend
3. Expected sign or trend
4. Order of magnitude
5. Bounds
6. Full probability density function (i.e. full quantitative risk assessment possible)

The scale provides a range from a qualitative indication i.e. whether it is good or bad for health, a rough estimate of the order of magnitude (i.e. 'hundreds of cases' of disease versus 'thousands of cases'), or a detailed risk-based assessment.

The questions covered the following categories of uncertainties:

- The climate system, e.g. heat wave frequencies and durations.
- The biological systems, e.g. the relationship between climate and insect distributions, and infection biology.
- The human systems, e.g. autonomous adaptation and responses of health systems, effectiveness of hygiene regulations, and disaster response.

The uncertainty typology or the 'Level of Precision' scale used is shown in the box 'Example of handling uncertainty'. The 'Level of Precision' question was relatively broad. Potentially, some participants could have scored health effects based on standard climate projections (e.g. the Dutch KNMI or global IPCC scenarios), while others could have assumed a broader ignorance regarding local climatic changes. Because the reasoning focused almost exclusively on uncertainties in assessing health impacts (i.e. translating a climatic change into its health impacts),

Health effects have:	Low policy-relevance	High policy-relevance
High level of precision health risk assessment	Tailored, prediction-based strategies (e.g. risk approach)	Tailored, prediction-based strategies (e.g. risk approach)
	Focus: low costs/efforts or co-benefits. **Example:** providing shelter for homeless people during cold spells.	Consider (but critically reflect on) costly and extensive options. **Example:** financing/subsidizing air-conditioning or other (advanced) cooling systems in buildings.
Low level of precision health risk assessment	Enhance system's capability of dealing with changes, uncertainties, and surprises (e.g. resilience approach).	Enhance system's capability of dealing with changes, uncertainties, and surprises (e.g. resilience approach).
	Focus: low costs/efforts or co-benefits. **Example:** general improvement in health care including research, and regular impact & adaptation assessments.	Consider (but critically reflect on) costly and extensive options, including precautionary options. Assess overinvestment risks and flexibility. Under which circumstances would "robust" measures be advocated and which? **Example:** changing building materials and increasing urban water and parks to reduce the impact of heat in urban areas.

Fig. 4.16 Appropriate adaptation approaches, considering uncertainty and policy-relevance of health effects (Wardekker et al. 2012)

rather than on climatic uncertainties, the scores were interpreted as 'given a climate scenario'. The individual scores, the expertise-weighted descriptive statistics, and the reasoning given for each score were assessed.

The second part of the study dealt with:

- The relevance of health effects to adaptation policy (e.g. where there are high health impacts, high societal or political salience, etc.),
- Specific uncertainties not mentioned in the reasons given for the 'Level of Precision' scores, and
- Uncertainty-robust adaptation options and strategies.

The relevance of health effects to adaptation policy was assessed by asking participants to select and rank the five effects they considered the most important, interpreting relevance in a broad way, and giving reasons for their choices. This separated the highly relevant from the less relevant effects, and highlighted the different reasons for relevance. For example: current vulnerability to the effect (heat-related mortality); large potential health and societal impacts, difficult to adapt to, and public fright factors (vector-borne diseases); and a large number of people affected and large potential economic impact (hay fever).

The implications of uncertainties for adaptation were discussed using various characteristics of policy options (e.g. costs, flexibility, encroachment, prediction versus capacity-enhancement). The results of this approach are summarised in Fig. 4.16.

Effect of Uncertainty on Decision–Making

> *"The uncertainty typology can be a very useful assessment tool for the selection and prioritisation of preferred climate adaptation policy in practice."*

The uncertainties assessed had a notable influence on the policy assessments conducted by the PBL for the Dutch government; it affected how they discussed climate change impacts on health and adaptation to these impacts. It became clear that adaptation in the health sector requires a strong focus on enhancing system resilience and on capacity building. The use of uncertainty typologies was also important; they allowed for a systematic and structured analysis of the uncertainties, distilling policy-relevant uncertainty information from the complex mix of imperfect evidence. They have led to the advice that a different policy approach would be needed, for example, for vector-borne diseases than for heat-related deaths. In effect they have made the various potential health impacts and their uncertainties comparable, which in turn have enabled adaptation strategies to be differentiated.

The typologies helped to focus on the most appropriate policy strategies, given the characteristics of both health impacts and policy options:

- For possible climate related health impacts characterised by ignorance, the most appropriate adaptation policies are those that focus on enhancing the capability of the health system and society in general in dealing with possible future changes, uncertainties and surprises e.g. through resilience, flexibility, and adaptive capacity.
- For climate related health effects for which rough risk estimates are available, 'robust decision-making' is recommended.
- For climate related health impacts which are less uncertain, tailored and prediction-based approaches are most appropriate.

By providing an interpretative framework for a complex mix of uncertain evidence, a systematic, rather than ad-hoc, formulation of policy advice is created. An example is the central role that uncertainties and uncertainty-proofing policy played in the workshop "Policy options for climate change and health" (PBL & WHO Europe, co-organised by the University of Utrecht, at the WHO office in Bonn, Germany, 11–12 January 2010). The outcome of this case has also been used in a recent follow-up of the PBL outlook studies on climate-proofing in the Netherlands to support the current national Delta Programme (addressing flood risks, fresh water availability, and urban stress). The developed framework for systematically dealing with uncertainties will be used to advocate a second Delta Programme, including a detailed health adaptation policy.

Authors: Arjan Wardekker, Jeroen van der Sluijs

Links to more information:
Wardekker, J.A., A. de Jong, L. van Bree, W.C. Turkenburg, and J.P. van der Sluijs (2012). Health risks of climate change: An assessment of uncertainties and its

implications for adaptation policies. *Environmental Health* 11: 67. http://www.ehjournal.net/content/11/1/67

The paper was summarized in the European Commission newsletter Science for Environment Policy: http://ec.europa.eu/environment/integration/research/newsalert/pdf/317na5.pdf

WHO and PBL (2010). "Policy options for climate change and health: Report on a joint WHO-PBL technical meeting". World Health Organization (WHO) Regional Office for Europe, and Netherlands Environmental Assessment Agency (PBL), Bonn/Bilthoven. http://www.pbl.nl/sites/default/files/cms/publicaties/pbl2010-who-pbl-technical-meeting-climate-change-and-health_0.pdf

Contact details: arjan.wardekker@gmail.com, tel: +31 70 340 7021; j.p.vandersluijs@uu.nl, tel: +31 30 253 7631

4.2.6 Flood Risk in Ireland

> **Country:** Ireland
> **Sector:**
> **Scale:** National
> **Organisation:** Public
> **Decision-type:** Strategic

Key Messages

The aim of this study was to look at how climate change has been integrated into existing policies for flood protection works and how robust those policies are under a range of climate change scenarios.

Key messages were:

- Reinforcement of the emerging picture that there is uncertainty in projections.
- Consideration of the performance of adaptation options over a wide range of uncertainty to ensure the robustness of the decision.
- The importance of communicating uncertainties in future projections so that decisions can be based on the full range of available information.

Background

In recent years flooding in Ireland has been quite extensive with substantial social impact. This case study looked at how climate change has been integrated into existing policies for flood protection works, and how robust those policies are.

The project was initiated by the Department of Geography at the National University of Ireland Maynooth and funded by the Science Foundation Ireland (SFI). The main beneficiary of the project was the Office of Public Works (OPW), the national agency responsible for flood risk reduction, whose policies were selected for review. Their policy reports have been influential in past decisions and they are one of the leading national agencies in Ireland that are climate sensitive and trying to accommodate changes.

Most of the work in flood defence in the past has been based on high resolution regional circulation models (RCMs), with a tendency to neglect other uncertainties such as those arising from the use of different general circulation models (GCMs), downscaling techniques, different socio-economic, emissions and land-use/soil sealing scenarios, and impact models. It is critical, for example to include results from a large sample of GCMs to assess the robustness of adaptation schemes. There is also a risk of overconfidence in projections due to the high resolution of RCMs. In adapting to an uncertain future it is important that more effort is made to capture the full range of uncertainties so that decisions are based on as much information as possible.

Process

The first step was to review the policy documents from the OPW. Identified safety margins incorporating climate change allowances were stress-tested using climate projections extracted for the Irish grid cell and pattern scaled to local catchments.

Climate data sources

- IPCC AR4 full range of GCMs (17 in total)
- Three emissions scenarios
- Time horizons 2020s, 2050s and 2080s

Fifty one climate projections were generated from IPCC AR4 data using the entire range of GCMs and three IPCC emissions scenarios. Change factors based on current climate conditions were determined and run through a weather generator to derive catchment scale information. This was then used to force a suite of hydrological models for four case study catchments. The model structure and parameter uncertainty of the hydrological models were accounted for and the sensitivity of safety margins for flood defences was assessed using risk response surfaces.

The OPW was involved in the study through informal meetings and conferences.

Uncertainty Assessment

The primary aim of the project was to test a set of adaptation options on flood risk for their robustness. This was done using sensitivity analysis on the flood defence thresholds incorporated in the policies. Peak flow safety margins of 20 %, for a medium emissions scenario, and 30 %, for a higher scenario were identified for new design flood defences, so sensitivity analysis was used to check how robust those

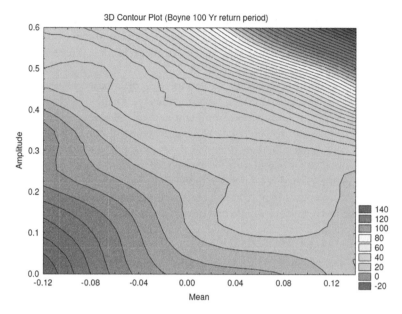

Fig. 4.17 Risk response surface for safety margins of 20 %. Only in case of a combination of relatively high mean precipitation change and high amplitude of precipitation a safety margin of 20 % will not be sufficient for the majority of projected changes in flooding

margins were over as much of the uncertainty range as possible. The research found that the performance of these safety margins differs between catchments. In some instances they were sufficient to cope with the range of scenarios analysed. In others, the safety margins were found to be too conservative for the range of climate projections considered, leaving high residual risk.

The project dealt with the following uncertainties:

- Emission scenarios,
- Global climate models,
- Natural variability,
- Hydrological model – both model structure and parameter uncertainty,
- Potential for future surprises in climatic conditions.

These uncertainties were dealt with in the following ways:

- Sensitivity analyses of which the results are displayed in risk response surfaces
- Risk response surfaces (see Fig. 4.17). These were used to visualise the effectiveness of the policy decision, given certain ranges in temperature and precipitation and the safety margins applied.
- Wild cards

Figure 4.17 displays the results of the sensitivity analysis in a response surface. Future precipitation changes are represented here as the mean and amplitude of the range of precipitation changes. It can be seen that a 20 % safety margin (based on current norms) shown as green area accounts for the majority of projected changes in flooding. However, it is apparent from the yellow and red areas (which exceed the 20 % allowance) that approximately one quarter of all simulations are not catered for by this safety margin. This can be thought of as the amount of residual risk associated with the policy of a 20 % allowance in flood design. The risk response surface was communicated to stakeholders at national meetings and conferences.

Following previous work done by others, particularly in the UK, the expansion of the sensitivity range on both the upper side and lower side to account for new extreme precipitation scenarios was also reviewed.

The project also considered uncertainty in the impacts models, i.e. the simple rainfall runoff models. This was done by looking at different model structures and parameter uncertainty.

Effect of Uncertainty on Decision-Making

"Ensure decisions are robust"

Using 51 different climate scenarios combined with uncertainties in downscaling and hydrological models, meant this was the biggest assessment of uncertainty in hydrological studies so far in Ireland. Previously the OPW has tended to use three scenarios to inform their decisions, but this work has reinforced their growing understanding that uncertainties need to be fully understood in order to take robust decisions. The OPW is moving away from a deterministic approach to adaptation decisions. This revolved around making specific assumptions about the way the climate will change, and designing structural engineering solutions such as building flood defences, perhaps with the capacity to increase their height in the future. They are now approaching decisions with softer techniques to ensure that they are robust under the full range of uncertainties involved.

A good example is Cork City, where a complete structural protection scheme against both fluvial and coastal flooding would have cost in the order of €140 m but would have given a reducing standard of protection over time. This is due to the fact that typical engineering approaches are built to a specific standard. As climate changes, the level of protection offered decreases potentially making the initial outlay of costs unjustified.

The proposed solution is therefore to provide partial defences through the city, with potential amendments to the reservoir operations and some localised protection works upstream of Cork, where land would be deliberately flooded to reduce fluvial flood risk. Barrages are also being considered as suitable alternatives to traditional defences (Fig. 4.18).

Fig. 4.18 Flood problems in Cork (Courtesy: Irish Examiner)

Author: Conor Murphy

Links to more information: Bastola, S., C. Murphy, and J. Sweeney. 2011. The sensitivity of fluvial flood risk in Irish catchments to the range of IPCC AR4 climate change scenarios. *Science of the Total Environment* 409(24): 5403–5415.

Contact details: conor.murphy@nuim.ie, Tel: +353 1 7083494

4.2.7 Coastal Flooding and Erosion in South West France

Country: France
Sector:
Scale: Local
Organisation: Public
Decision-type: Operational

4 Showcasing Practitioners' Experiences

Key Message

This project deals with the increased risk of coastal flooding and erosion through sea-level rise in South West France.

The key message from the project was:

- Using a low/no regret approach serves many functions, such as solving the flood problem, adding value to natural reserves and creating new potential for recreation.
- Add other messages, such as the feasibility of taking meaningful action in the absence of precise predictions of future changes, etc.
- Meaningful coastal investments can be made in the absence of precise predictions of future changes.
- Climate change impacts can be strong drivers to implement projects that strive for both current and future vulnerability.

Background

The lido[5] between Sète and Marseillan in the Languedoc-Roussillon region of France was threatened by sea level rise and erosion. During the last two decades coastal erosion and flooding have caused increasing traffic disruption on the road between the two towns and the inland biodiversity and heritage was additionally impacted by storm surges. Protection was also needed for economic activities such as vineyards and oyster farming in the Thau pond, as well as the sand beach and the local campsite.

The threat triggered a comprehensive spatial planning project run by the Community of Communes. The project was driven by a desire to counter beach erosion and the climate change dimension wasn't initially considered; it was launched in 2000 with a view to targeting soft protection measures rather than concrete devices. Sea level rise was primarily considered during the implementation phase to ensure that the measures taken would be sustainable in the long term.

Funding was provided by the State, the local authorities (regional and departmental) and the European Union through the European Regional Development Fund (ERDF). An Interreg III project has also been conducted for its demonstrative and innovative purposes.

Process

A study into the feasibility of moving the road, and the sustainable land planning of the lido, started in 2003 and was completed in 2005 with many public consultations. The public consultation is a mandatory process in France, required for significant

---
[5] Public place for beach recreation, including a pool for swimming or water sports.

spatial planning projects in order to identify natural, social and cultural issues. After completion of the consultations, the development project was finalised and the works started in early 2007.

> **Climate data used**
>
> Ministry of Environment recommendation on sea level rise for long-term planning to be +25 cm by 2050 (DGEC/ONERC 2010).

The current vulnerability to flooding was well known, but data from the Ministry of Environment recommended considering a sea level rise of over 25 cm by 2050. The Community considered the option of leaving the road as it was, but the cost-benefit analysis delivered many benefits of a strategic relocation of the road behind the lido. One of these benefits was the fact that such a move, combined with a regeneration of the sand dunes would "climate-proof" the area against potential flooding for over 50 years. The new road became operational during summer 2010 and the rehabilitation of the sand dunes of the lido continued until 2011 (Figs. 4.19–4.21).

Fig. 4.19 Recurrent erosion impacts on the coastal road

Fig. 4.20 Global overview before the commencement of the project showing the road situated next to the beach

Fig. 4.21 Global overview after completion of the project showing the road moved inland and the restoration of a wider beach and sand dune

Uncertainty Assessment

The two main types of uncertainty were:

- The exact value of sea level rise and its associated extreme wave heights from storm surges.
- Erosion trends under sea level rise.

To cope with the uncertainty surrounding the magnitude of sea level rise, the project decided to combine the relocation of the road with protection of the sand ridge and restoration of the beach width. Expert advice from the technical advising contractor was taken and there was public consultation with stakeholders. The road was moved behind the lido and the sand dunes restored to a height of 4.2 m above sea level. The new road relocated inland has been raised by 1.5 m in order to reduce the risk of permanent road flooding during strong storm surges and to anticipate the new flood risk management scheme; the regional Disaster Risk Management unit has strongly supported the idea of raising the road.

In addition, the restored dunes were populated with plants stored prior to the start of the project and the position of the dunes is now being monitored with cameras along the beach line. Some innovative coastal defense measures are being taken (e.g. sunken geotubes[6]) to attempt to minimise the effects of erosion, and these are also being monitored. This multi-measure approach provided good resilience to the rising sea level and is "low regret" in the sense that the adaptations provide other benefits such as recreational facilities and Natura 2000 sustainability.

Effect of Uncertainty on Decision-Making

"Time is needed to convince a community that changes should be sustainable"

The project did not evolve exclusively from a need to consider climate change, but impacts related to sea-level rise, such as erosion and flooding, were key drivers. The Community of Communes wanted a long-term solution to the problems and found that the best way was to produce defences high enough to deal with all eventualities. This solution was a "low regret" solution as it also provided biodiversity, economic and recreational benefits. Exchanges between the project leader, expert and the regional DRM unit have helped to consider sea level rise in a pragmatic way.

The Community of Communes has been able to propose an amended solution to the local problem. Dunes were previously considered obstacles to the development

---

[6] The geotubes are sediment-filled sleeves of geotextile fabric and used to build structures such as breakwaters, shoreline protection or island creation.

of tourism and at the beginning of the project some decision-makers just wanted to build dykes to keep the sea at bay. The proposed solution has restored the beach and helped sustain the local economic activity. It also provides the necessary protection from erosion and flooding.

Author: Bertrand Reysset

Links to more information: http://www.thau-agglo.fr/IMG/pdf/Dossier_Presse_Lido_2011-2-2.pdf, http://www.developpement-durable.gouv.fr/IMG/pdf/ONERC_lettre_2.pdf

Data sources: DGEC/ONERC (2010), *Prise en compte de l'élévation du niveau de la mer en vue de l'estimation des impacts du changement climatique et des mesures d'adaptation possibles,* Synthèse n°2, 6 p. http://www.developpement-durable.gouv.fr/IMG/pdf/synth_niveau_mer.pdf

Contact details: bertrand.reysset@developpement-durable.gouv.fr, tel: +33 1 40 81 92 94, c.cazes@thau-agglo.fr, webredac@thau-agglo.fr

Thau agglo, 4, avenue d'Aigues, BP 600, F- 34110 FRONTIGNAN cedex, Tél. 04 67 46 47 48/Fax. 04 67 46 47 47

4.2.8 Québec Hydro-Electric Power

Key Messages

> **Country:** Canada
> **Sector:**
> **Scale:** Regional
> **Organisation:** Public (State-owned)
> **Decision-type:** Strategic

This case study was designed to determine whether climate change should be taken into consideration when developing a hydro-electric power plant refurbishment strategy.

Key messages from this project were:

- The realisation by the hydropower company that there was no such thing as a single "best (climate change) scenario" and that multiple scenarios should be used to deal with climate change uncertainties.
- Clear communication between the climate scenario developers and the operation management and openness to mutual knowledge transfer were most important in the outcome of the project.

Fig. 4.22 Manic 2 Power House on the Manicouagan River (Source: Hydro-Québec)

Background

> *"There is no such thing as a single "best" scenario in climate change"*

After several decades of operation, a number of dams and hydropower stations of the state owned company, **Hydro-Québec** needed refurbishment (Fig. 4.22). Changes in climate have already and will further affect the flow regimes of the dammed catchments. For example, until now winter precipitation has largely been snow, but this is now changing to include rain which ideally needs to be harnessed.

Hydro-Québec, was the primary stakeholder of this project. Their research division, IREQ (Institut de recherche d'Hydro-Québec), conducts research into energy related fields including the assessment of climate change impacts on the watersheds of their power generation stations. However this time it was the operation management who took the step to request concrete climate change information.

The company wished to update its generating equipment to provide state of the art facilities. As part of this process it wanted to evaluate future hydrological conditions to determine their effect on plans for renovation. If they established that climate change was likely to affect their long-term decisions, they planned to carry out more

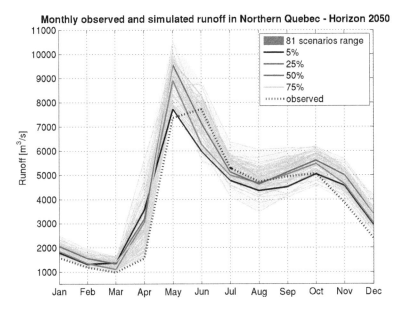

Fig. 4.23 Annual cycle of observed and simulated runoff in a northern Québec watershed. The presently observed runoff is shown as the dashed line. The four selected future scenarios representing the 5th, 25th, 50th and 75th percentile of the range of projected change are shown in colour over the range of all scenarios used. The selection was based on cluster analysis of multiple indicators critical in dam operation and management

in-depth studies of the impacts for specific catchments and sites to be modernised. Their initial approach was to base their study on the "best (climate change) scenario". However, following involvement in meetings and workshops it eventually became clear to them that climate system and projection uncertainty cannot be considered using a single scenario. A sound approach was then developed to review climate change effects under a broader range of conditions. In the end the economic impact study utilised four different future hydrological scenarios.

Process

Initially, a request was made to the Ouranos Consortium, a private, non-profit making organisation advising in the areas of climate sciences, impacts and adaptation, for the "best climate change scenario" to help the company with their plans for plant refurbishment. This resulted in an investigation into climate simulation data and their hydrological impacts and after many meetings and exchanges about the needs of the stakeholder, four projections, representing the 5th, 25th, 50th and 75th percentiles of the range in uncertainty were asked for by the client and as such provided. This is demonstrated in Fig. 4.23 showing changes in runoff.

The work was shared between the Ouranos Consortium who produced the climate scenarios and IREQ who did the hydrological modelling. Clear communication and

> **Climate data used**
>
> - 81 climate simulations composed from:
>   - 73 global climate models from CMIP3 (scale approx. 250×250 km)
>   - 8 regional climate models from Ouranos CRCM4 simulations (scale 45×45 km)
> - Climate variables used to drive a hydrological model: daily precipitation, minimum and maximum temperatures

openness to mutual knowledge transfer were key to the results. For the production of hydropower, precipitation in combination with temperatures is the key climate vulnerability. The meteorological variables were transformed into stream flow using a hydrological model and the four percentiles described above were selected to cover the uncertainty. The final economic evaluation was done by Hydro-Québec in order for them to decide if there was enough change to affect their investment in infrastructure.

A short description of the study was presented outlining the general impacts of climate change on hydrology in the north of Québec.

Risks for hydropower production under different future hydro-climatic conditions include a loss of efficiency of old installations and possible complications in the management of the available water. For example, a release of excess water in the reservoirs would mean a loss of hydropower production. In refurbishing their installations, Hydro-Québec was trying to cope with these vulnerabilities and risks.

> **Example of handling uncertainty: Multi-criteria cluster analysis**
>
> An ensemble of 81 climate simulations was analysed for 11 watersheds. Daily values for each watershed were bias corrected and used to drive a hydrological model to obtain future stream flow scenarios. They were then filtered in a multi-criteria cluster analysis to represent the 5th, 25th, 50th and 75th percentiles of the range of uncertainty in the hydro-climatological projections. Cost-benefit analyses were then performed using these four different hydrological scenarios. In this manner the range from 5 to 75 % (=70 %) of the uncertainty was effectively addressed.

Uncertainty Assessment

The uncertainties taken into account in this study included:

- GHG emission scenario uncertainty,
- Climate model uncertainty,
- Climate system uncertainty,
- Regionalization uncertainty.

Different possible developments of future societies were accounted for by using three GHG emission scenarios in the climate simulations ensemble. Climate model and climate system uncertainty were addressed by including multiple simulations from 16 different global climate models and one regional climate model. Uncertainty of regionalisation of the scenarios was accounted for by using four different empirical downscaling methods in the production of regional hydrological scenarios.

The methods used to analyse the different types of uncertainties were as follows:

- Project scenario analysis (see box),
- Expert elicitation through consultation with the Atmospheric Sciences department at Université du Québec à Montréal,
- Sensitivity analysis of bias correction methods/empirical downscaling,
- Multi-model ensemble using the maximum number of models possible,
- Stakeholder involvement between parties at Hydro-Quebec and Ouranos.

Example of handling uncertainty: Project scenario analysis

Eleven different watersheds had to be identified and analysed. In some cases watershed boundaries had to be re-examined in order to be correctly modelled and to obtain optimal observational data for the empirical downscaling. These iterations were needed to set up the physical description of the problem. Then, the options of covering uncertainty using different numbers of scenarios were played through to demonstrate that the request of "the best scenario" might be over simplified.

By employing exclusively Hydro Québec's operational hydrological model, the uncertainty from hydrological model choice could not be considered. This would require a hydrological model ensemble. Likewise, it was beyond the scope of this study to relate the magnitudes of uncertainty from climate change projections to those from cost-benefit analysis. Both issues are important but relatively new fields of research and shall be addressed in subsequent, more detailed assessment.

Effect of Uncertainty on Decision–Making

Uncertainty has had a profound effect on the course of this study, commencing with the realisation that more than one climate change scenario needed to be taken into account.

The four selected scenarios were used as varying assumptions for a cost-benefit analysis to assess the impacts of increased runoff on hydro-power assets. Based on the results of this analysis the stakeholder has decided that the impacts of climate change are of a magnitude that need to be taken into account in the planning of renovations of hydropower facilities. Thus, more in depth studies of climate change impacts will be conducted and Hydro-Québec will be reviewing its position in more detail to achieve a clear picture of cost-benefit options due to climate change impacts.

Authors: Marco Braun (Ouranos), René Roy (IREQ) and Diane Chaumont (Ouranos)

Links to more information: http://www.ouranos.ca, http://www.hydroquebec.com/en

Contact details: braun.marco@ouranos.ca, tel: +1 514 282 6464 306

4.2.9 Austrian Federal Railways

> **Country:** Austria
> **Sector:**
> **Scale:** National
> **Organisation:** Public
> **Decision-type:** Operational+Strategic

Key Messages

"Give information to those who need it"

This case study focused on adaptation in railway infrastructure and how uncertainties in future climate need to be properly considered when time-scales of 100 years are involved.

The key messages are:

- Trend analysis is a useful way to handle uncertainties.
- Constant feedback between company staff and experts is necessary throughout the process.
- Messages must be communicated clearly and in a language which matches the stakeholders language, particularly concerning uncertainties.
- Climate change is usually just another uncertain issue amongst others that companies have to handle traditionally.

Background

"Try to be practical"

The **Austrian Federal Railways** (ÖBB – Österreichische Bundesbahnen) runs the national railway system of Austria. It is entirely owned by the Republic of Austria and is divided into several separate businesses that manage the infrastructure and

4 Showcasing Practitioners' Experiences

Fig. 4.24 Winter service ÖBB – West part of Austria in January 2012 (Photos: ÖBB)

operate passenger and freight services. Since 2003 it has also run Austria's largest bus company with its intercity networks (Fig. 4.24).

The ÖBB is a significant organisation, carrying about 450 million passengers a year. It has about 4,800 km of route network and more than 1,000 railway stations. Given the long life-span of up to 100 years in investments in major transport routes, bridges, tunnels etc. the ÖBB recognised the importance of properly considering changes in future climate when making decisions. After all, the company knows only too well that there is little tolerance from passengers towards the late running of trains.

In 2010, the company contracted the Austrian Environment Agency to help identify potential climate change impacts on rail infrastructure and develop recommendations for adaptation. The aim was to investigate as many meteorological variables and climatic changes as possible that might have an impact on the company's infrastructure and security of service. The company wanted to find practical solutions for problems, whilst taking into account the best scientific knowledge available. The ultimate goal was to incorporate the findings in the company's long-term risk strategy.

Process

*"Maintain constant feedback throughout
to achieve a robust outcome"*

The company was not new to the concept of uncertainty, partly because Austria is an alpine country and used to natural hazard management. They had realised uncertainty is not exclusive to climate change and already affects current decisions in natural hazard management.

Senior executives and company experts in the fields of research and innovation, natural hazards and sustainability were assembled into a steering group and included in every step of the project. Such continuous involvement by company staff in the project was seen as critical to its success. Experts from the Institute of Meteorology at the University of Applied Life Science were part of the project team, also participating in the steering group.

The project focused primarily on climate related risks and the company representatives were generally open and very interested in such matters, being aware of the impact that weather related events can have.

The steering group met approximately every 2 months and this close cooperation between experts with vital information was important to the success of the project. Three workshops were also held to involve other members of the company and discuss the following topics:

- Climate change impacts on railway infrastructure – discussing the overview table.
- Vulnerabilities with specific focus on natural hazards – using trend analysis from company data. It was during this discussion that concerns about uncertainties were addressed with one stakeholder declaring *"You can't tell us what will happen in 2020 in region xxx, so how should we know what to do about this?"* The company's pragmatic answer to this was to provide clear guidance to staff required to implement decisions.
- Climate change adaptation options – dedicated to presenting possible options for the future and getting feedback from the stakeholders.

Climate data used

Regionalised climate scenario were based on ECHAM5 and HADCM3 models and A1B and B1 IPCC GHG scenarios

---

The first step was to produce an overview table on observed climate impacts for railway infrastructure and some operational issues. This was based on qualitative information stemming from research projects, grey literature and other information sources, and was used as the first basis for the discussion with company representatives. Past observations and stakeholder knowledge were combined with expert judgements using regional climate data so that important climate related impacts and trends could be identified for the ÖBB. In addition, past trends were extracted from company data to see if there were links between disturbances to operations and meteorological events (see Fig. 4.25).

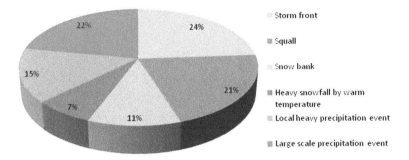

Fig. 4.25 Disturbance cases between 1990 and 2011 clustered by meteorological events (ÖBB data analysed by H. Formayer)

Uncertainty Assessment

It became obvious during the course of the project that dealing with the following uncertainties were key for a good and robust result:

- Uncertainties inherent in climate scenarios (emission scenarios, global models, regional scale issues, problems with consistency of data series). These were dealt with by involving an expert climate meteorologist and working with trend analysis.
- Changes in method of data selection and documentation in the ÖBB internal database on past natural hazards which were used for the trend analysis. Sensitivity analysis was applied to this data.
- No regret/low regret analysis: The Environment Agency collected adaptation options from the literature and highlighted if these options were no-regret or low-regret. The list was discussed with the company's staff to understand if the options would benefit the company and if they could be connected with already existing measures. Considering uncertainties involved, the flexibility of the options was assessed as well.

Handling uncertainty – Trend analysis

More than 1,000 events over the previous 20 years were analysed and compared to parameters such as heavy precipitation, high winds or excessive temperatures responsible for causing disturbances. This formed the basis for the vulnerability assessment and the determination of future trends, although there was some concern over the integrity of this database. Future trends in climate parameters and thus impacts on infrastructure (e.g. rail buckling, infrastructure damage due to floods, storms or heavy snow fall) were then determined based on available regional climate models and expert knowledge.

Other methods of handling uncertainty included:

- Stakeholder involvement

Effect of Uncertainty on Decision–Making

"Implement now to avoid greater costs in the future"

*"Nobody knows what will really happen
so it is safer to act now"*

The project had two very positive outcomes. Firstly, future investment will be climate-proofed; due to the uncertainties in future climate projections, it was decided that planning new infrastructure should not focus on one single "optimal" solution but should be made more robust by taking into account a range of possible climatic changes. Thus, in the case of transport infrastructure, multiple-benefits, no-regret and low-regret adaptation options were recommended.

One example is that of future track drainage. Trend analysis showed that in certain regions future rainfall may become more intense. To cater for this, track drainage will need to be improved. The company reviewed the range of likely outcomes and decided drainage should be improved in some regions to cover all likely eventualities.

Secondly, there was improved sensitivity to climate issues; having experienced the project process, company representatives have built climate change issues into their long-term strategy and developed a sound basis on which to consider such issues in the future.

Author: Andrea Prutsch

Links for more information: http://botany.uibk.ac.at/neophyten/download/09_OeBB_Rachoy_KLIWA.pdf, http://www.oebb.at/infrastruktur/__resources/llShowDoc.jsp?nodeId=29841913

Contact details: andrea.prutsch@umweltbundesamt.at, tel: +43 1 313 04 3462

4.2.10 Dresden Public Transport

Country: Germany
Sector:
Scale: Local
Organisation: Public
Decision-type: Strategic

4 Showcasing Practitioners' Experiences

Key Messages

This project helped refine the current business strategy of a public transport provider in Dresden, Germany to take into account the future effects of climate and demographic change.

Key messages are:

- New tools, such as fuzzy cognitive maps, help clarify uncertainties and identify appropriate strategies within an environment facing a complex mix of challenges.
- Company executives were stimulated to consider the implications of climate change amongst other uncertainties in their decisions.

Background

"An expert partner in the project is crucial"

Public transport is highly sensitive and vulnerable to external impacts which affect the complex relationship between infrastructure, technology, time schedules, and volatile customer behaviour. In a dynamic developing city, the public transport provider needs to deal with changing conditions. Uncertainty in investment funding from the public budget as well as the high dependency on political decisions means that constant planning and refinement of plans is needed.

Climate change primarily impacts this industry through extreme weather events; inherent uncertainties in these have a big influence on both the planning of infrastructure and daily operations of the business. For example, a major flood in 2002 caused roads to be closed and damage to infrastructure which had a long-term impact on the public transport system (Fig. 4.26). Then, in 2003, a heat wave with extreme high temperatures caused discomfort for customers and drivers in buses and trams without air conditioning. In addition, storms, heavy snow fall or ice on the overhead wire can disrupt operation or cause damage through fallen trees etc.

The main goal of the case study was to refine the company's business strategy in the face of future challenges such as climate and demographic change. The company has already taken action to adjust the time schedule of trams and buses in the winter season to handle the possible impacts of continuing snow fall. Economic and technological challenges, such as the increase of energy prices, have also been considered through the introduction of buses with hybrid technology.

Fig. 4.26 Impact of major flood in Dresden in 2002

The project was conducted as part of REGKLAM, an integrated regional climate change adaptation program. It is part of KLIMZUG financed by the German Federal Ministry of Education and Research, involving partners from politics, administration, business and science. The case study itself was run by Technische Universität Dresden (TUD) (under the lead of Chair of Environmental Management and Accounting) and involved the two CEOs of the company along with representatives from company departments such as strategic planning, financial control and human resources. There was considerable understanding within these departments of the negative impacts that climate change is having on the day-to-day running of the transport system through the increase in extreme weather events. The objective was to discuss the final results with the city government to plan for a resilient public transport system.

Process

Figure 4.27 presents the process that was used in the project to develop and use/transfer scenarios in an iterative way. The process began with a kick off meeting in August 2011 to determine the goals. Then, after some desk research, a number of workshops were held, first with experts and then with company representatives, to select key climatic and non-climatic challenges and to analyse the future

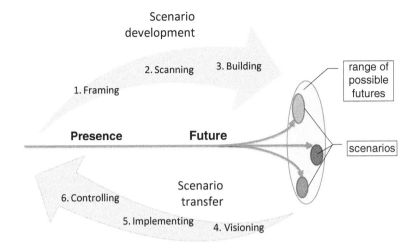

Fig. 4.27 Iterative development of scenarios

development of their associated uncertainties (e.g., climate or demographic change). Accordingly, up to three assumptions for the future development of the key challenges were defined. Various scenarios are developed from this by applying different assumptions to potential pictures of the future. These possible futures will be discussed in workshops with senior executives and options for adaptation identified. The project finished mid 2013.

Climate data sources

- Historic data from the Met Office
- Forecast data using climate models WEREX IV, REMO, CLM and WETTREG (Met Office)
- IPCC emission scenarios

As part of REGKLAM, data was taken from fact sheets developed by the chair of meteorology of TUD. These gave historic data for two time periods up to 2005 for important regional and local climate parameters such as average temperature, average precipitation, dry and hot weather days. They also provided ranges of forecast data for two further time slices up to 2100.

From discussions with company executives however, it became clear that interest was particularly focussed on extreme weather events as these are likely to have the biggest impact on the business. Information was taken from the literature and the whole business environment was scanned. In a first step all potential challenges – 60 in total – were identified and categorised. These were reduced to 19 which particularly affect this public transport sector in order to tackle the problem.

Uncertainty Assessment

Uncertainty in dealing with extreme weather events exists to the extent that no assumptions or prognoses can be made for their future occurrence. The meteorologists in the project developed prognoses for average temperature and precipitation, but they were not able to make such "assumptions" for the occurrence and impact of extreme weather events. The uncertainty related to incomplete knowledge of such events on business challenges was therefore addressed through the use of Fuzzy Cognitive Mapping.

Nineteen climatic and other business challenges were identified in workshops with the stakeholder using Fuzzy Cognitive Maps (Fig. 4.28), with some of the influences described in full below. Possible relationships between the influence factors were identified and assessed according to the strength of the influence. For example it can be seen that extreme weather events such as heavy precipitation, floods, heat waves etc. (EXTWE) have a great influence on the development of information/communication/distribution systems (ICDSY).

Influence factors that have a significant effect or are highly affected by others within the whole system were selected as major key challenges for the next step in the process. Examples included an increase of extreme weather events, changes of customer behaviour, an increase in the development of technologies, and increasing political influence. In this way important relationships between factors affecting a business are identified and the uncertainties are reduced by dealing with these complexities.

The company felt that, through the use of the fuzzy cognitive map, the project provides a clear view on the connections between all factors that influence their business and on the possible effects of their decisions. They feel that it will ease their selections between different options for decision making.

Other methods of handling uncertainty were as follows:

- Scenario analysis ("surprise-free"),
- Expert elicitation,
- Sensitivity analysis,
- Stakeholder involvement,
- Wild cards/surprise scenarios.

Effect of Uncertainty on Decision-Making

> *"Time is needed within the process to pause and reflect"*

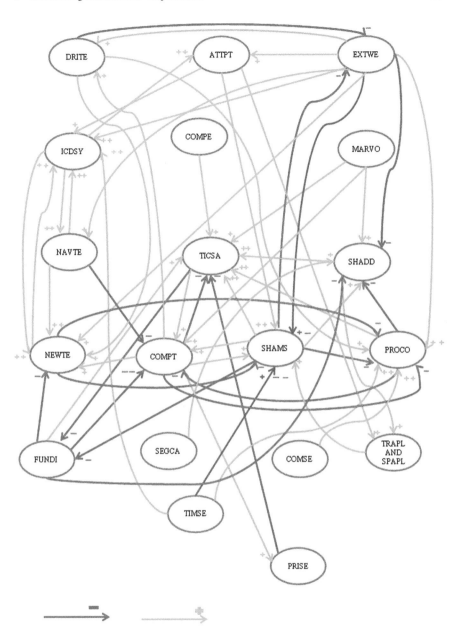

Fig. 4.28 Fuzzy Cognitive Map indicating relationships between influencing factors. *Green arrows* show positive influences and red negative ones. *Blue colour* stands for a relationship that can be both positive and negative. Fuzzy Cognitive Map influences: *EXTWE* Occurrence of extreme weather events, *TICSA* Ticket sales and revenues, *SHADD* Shareholder expectations of deficit development, *NEWTE* New technologies, *COMPT* competition within the public transport community, *SHAMS* share in the modal split, *PROCO* procurement cost, *ICDSY* information, communication, distribution systems, *FUNDI* Funding, *PRISE* Price sensitivity of customers, *TIMSE* Time sensitivity of customers, *COMSE* Comfort sensitivity of customers, *TRAPL* Traffic planning, *SPAPL* spatial planning, *SEGCA* Segregation of duties to the commissioning authority, *NAVTE* Navigation technologies, *MARVO* Market volume, *ATTPT* Attitudes/public transport supporters, *DRITE* Drive technologies or fuels, *COMPE* Compensation

By analysing the whole business environment and identifying the major future challenges, the managers and decision-makers were encouraged to think creatively. This led to a new view on existing strategies and actions and stimulated action to address the associated uncertainties.

The company is very aware that some issues will be strongly influenced by climate and climate change mitigation. For example, diesel engines will disappear in the future, but no-one can yet say what will replace them. Therefore, they need to be involved in the research process. The company culture demands that time is allocated to allow ideas, options and tools to become integrated into general practice. New methods and tools for strategic planning and long-term thinking were introduced and the end result will be an implementation plan for climate change adaptation measures.

Authors: Julian Meyr and Edeltraud Guenther

Links to more information:
For information on the institution leading the case study: http://tu-dresden.de/die_tu_dresden/fakultaeten/fakultaet_wirtschaftswissenschaften/bwl/bu/
For information on the background to the project: www.regklam.de

Contact details: ema@mailbox.tu-dresden.de

4.2.11 Hutt River Flood Management

Country: New Zealand
Sector:
Scale: Local
Organisation: Public
Decision-type: Strategic

Key Messages

> *"Better to consider a full range of uncertainties now than to put off action until the future when costs will be higher"*

> *"Uncertainties cannot be dismissed as an area scientists don't understand"*

This project aimed to improve the understanding of flood risks under the uncertainties of a changing climate in a river basin in New Zealand.

4 Showcasing Practitioners' Experiences

Key messages from the project are:

- The traditional tendency to project historical experience forward is a poor strategy in an uncertain climate because the future is unlikely to be like the past.
- Studies of uncertainties can expose the limits of static flood protection and of emergency planning. Understanding this increased practitioners and community consideration of a wider range of options and adaptive management in space and over time.
- Simple models can be used to explore uncertainties at low cost.
- A workshop process helps increase awareness of uncertainties in future flood risk and their planning implications and influence responses.
- Visual depictions are a powerful way to communicate the effects of climate change uncertainties.

Background

The aims of the project were to:

- Find a simple and low cost method of characterising the effect of climate change on flood frequency across a range of possible futures, and
- Demonstrate whether this influenced understanding and responses to changing flood risk.

The traditional way of using best estimates as single numbers or averages mischaracterises the range (uncertainty) and especially damaging extremes, thus entrenching the perception that protection structures offer safety for long-lived settlements and infrastructure. The project highlighted residual risks to settlements above design flood levels which increase with climate change. It was applied to the Hutt River basin, assessing flood frequency and potential damages of increased inundation levels with climate change. The project was run by the New Zealand Climate Change Research Institute at the Victoria University of Wellington, funded by the government Ministry of Science and Innovation. The primary stakeholders were the Greater Wellington Regional Council and Hutt City Council.

Flood risk is enhanced by climate change and there are substantial risks to urban communities which vary according to socio-economic status and ethnicity. Current methods used in flood risk management in New Zealand do not account for the effects of climate change on flood frequency and in particular, do not consider extremes which represent the uncertainties across the range of future changes. Until now, councils have taken a static, inflexible approach to climate risk in their flood management which has had the effect of entrenching and exacerbating this risk. In addition, averages and single scenarios are often used which underestimate extremes. Consequently, design flood levels used for flood risk management can result in inadequate protection for changing climate risk and give rise to a false sense of security to decision-makers and their communities. A more nuanced, risk-based approach to the effect of changing climate on flood frequency requires consideration of a wide range of alternative scenarios, but this is often constrained by the high cost and complexity of modelling. This project illustrates a simplified approach for evaluating uncertainty in future changes in flood frequencies based on different climate change scenarios, using the Hutt River in New Zealand's lower North Island.

Process

The case study comprised three parts:

- Modelling the effect of climate change on the Hutt River flood frequency and the potential damages from resulting inundation,
- A survey of households on how they responded to flood risk and their views on future climate change induced flood risk,
- A workshop with practitioners across a number of councils in the Wellington region and follow-up interviews with a sample of them.

Climate data sources

- Historical flood data (1972–2008)
- 12 GCMs, statistically downscaled
- Four different emissions scenarios
- An algorithm to infer changes in extreme rainfall based on changes in monthly mean climate

---

The model used 48 downscaled scenarios to derive changes in monthly average rainfall and temperature in the Hutt river catchment. From these, a simple algorithm determined changes in extreme rainfall which were run through a hydrological model calibrated to the Hutt River.

The results were tested at the workshop to gauge how the participants would respond. Participants included local government practitioners across strategic planning, urban planning, engineering, hazards management scientists, emergency management, and flood management, being those most involved in decision-making on flood risk. The uncertainties were presented visually as a changing risk. This increased the awareness of the participants to a range of possible futures, especially the damage consequences at the extremes, and the need for them to consider a wider range of more flexible responses. They realised that considering the uncertainties more transparently could potentially affect the design and planning assumptions over the life of the flood protection structures. This could thus reduce the risk to the people and assets currently protected. Presenting the dynamic nature of the risk in descriptive and visual form focused the thinking of the participants on the implications and their possible responses.

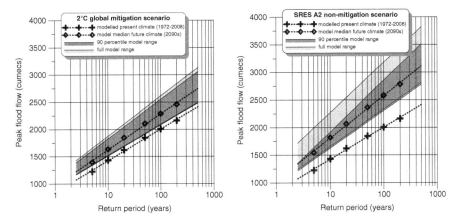

Fig. 4.29 Changes in exceedence probabilities under different emission scenarios. The *black dots* and *solid line* show estimated exceedence probabilities for a range of design flood volumes. The *dotted line* shows the flood volumes for alternative emissions scenarios in 2090 (*left*: 2 °C stabilisation; *right*: A2 SRES emissions) for a range of climate models. The *light grey* band shows the full model range, whereas the *dark grey* band shows the 10–90 percentile model range. A return period of 100 years in the *left hand graph* becomes 30 years and for the *right hand graph* becomes 20 years

The risk context of the visual presentation also resonated with elected councillors. A time and functional element to discussions was introduced, whereby the participants could identify activities with different lifetimes and conceive that changes could be staged over different timeframes to address the changing risk. This was effectively a discussion of adaptive management.

Uncertainty Assessment

The prime uncertainty addressed in this study was the effect of climate change on flood frequency, especially at the extremes. A quick and relatively low-cost methodology to explore the implications of alternative climate change scenarios for flood frequency was presented and applied in a stakeholder workshop setting. Exceedance probabilities, as shown in Fig. 4.29, appeared to increase under all scenarios but with considerable differences between alternative emissions scenarios and climate models. Understanding the full model range and how it changes in frequency emphasises the importance of low probability high impact events for planning and design of responses.

The approach used to assess the potential changes in flood frequency through to the 2090s comprised three steps:

- Statistically downscaled 12 GCMs and four emissions scenarios were used to produce 48 alternative climates (i.e. changes in *monthly average* rainfall and temperature) over the twenty-first century for the Hutt River catchment

- A simple procedure (algorithm) was used to estimate changes in *extreme rainfall* for the catchment
- Hourly rainfall data was run (both historical and adjusted for future climate changes in both means and extremes) through a hydrological model to derive flood frequencies under historical and 48 alternative future climates.

> **Stakeholder consideration of uncertainty**
>
> Flood frequency information affected by climate change was presented visually to participants from councils in the Wellington region. This resulted in participants questioning their reliance on flood warnings, emergency management and levees. The information focused attention on a wider range of complementary response options including protection, accommodation, spatial planning and retreat and the timing of different decisions.

The analysis represents a key advance on those earlier studies in that it quantifies uncertainties in the projected changes depending on emissions and climate models. This supports a more risk-based assessment of impacts and response options and avoids a premature collapse of a range of futures into single estimates, or reliance on simple scaling of current flood volumes that may not account for non-linearities and thresholds in catchment hydrology.

The following methods were used in combination for analysing uncertainty:

- Scenario analysis,
- Sensitivity analysis,
- Probabilistic multi-model ensemble,
- Stakeholder involvement.

Effect of Uncertainty on Decision-Making

"Studies such as these can increase a community's acceptance of a wider range of appropriate options"

This project has catalysed a shift in thinking from static safety and path dependency, to thinking about how to build flexibility into decision-making. For example, a realisation that the bottom of the Hutt catchment could face risks from increased runoff and rainfall, sea level rise, and storm surges, has led to a sharper focus on managed retreat as an option for one low-lying area. The Greater Wellington Regional Council, responsible for the Hutt river management, is including the findings of this study in a review of their flood risk management plan. They have also used the results to discuss a wider range of response options with the local council in the area of the Hutt valley.

Fig. 4.30 Flooding of the Hutt river

Modelling a range of possible futures and showing how a changing climate could affect flood frequency has enabled stakeholders to see the value of the approach developed for their consideration of future risk. Within the community there is an expectation of continuous structural protection. Examination of uncertainty however, exposed the limits of static protection and enabled practitioners to more seriously consider complementary measures that could address changes in climate impacts. These limits may include the costs of raising higher levees and of higher residual damage, as extreme events increase in frequency and intensity and design levels are exceeded. The need for continuous consideration of changing climate risk was also highlighted.

Feedback received from the local government organisations was very positive. They felt it gave them a framework to think about changing climate risk, allowing them to quickly scan responses and discuss them with the elected councillors and local urban councils to consider the implications for a range of options, their costs and timing to enable uncertainties to be a catalyst for decision-making for the future (Fig. 4.30).

Author: Judy Lawrence

Links to more information:

Reports from the research programme can be found here: http://www.victoria.ac.nz/sgees/research-centres/ccri/ccri-publications

The Ministry for the Environment Guidance on the effect of CC on flood flows and which includes the methodology that we used to generate the effect for the Hutt Valley can be found here: http://www.mfe.govt.nz/publications/climate/climate-change-effects-on-flood-flow/tools-estimating-effects-climate-change.pdf

Contact details: judy.lawrence@vuw.ac.nz, +64 (0)21 499011

4.2.12 Communication of Large Numbers of Climate Scenarios in Dutch Climate Adaptation Workshops

Country: Netherlands
Sector:
Scale: Regional/local
Organisation: Public
Decision-type: No decision

Key Messages

This study used workshops to discuss climate change impacts on spatial planning. Climate uncertainties were addressed by means of scenario analysis and different ways of visualising scenario outcomes were tested.

Key learning experiences are:

- The method of presentation of climate change scenario information is key to the understanding of decision-makers.
- Interactive forms of visualising scenario outcomes allow stakeholders to handle the data themselves and so to better understand the impact.
- Policy-makers have a tendency to focus on the 'middle of the road' scenario, whilst scientists focus on extremes, highlighting the inadequacy of a single scenario map.
- There is a high risk of using a single map as decision makers tend to see this as a prediction rather than a projection.
- The challenge of uncertainty combined with high costs of extreme adaptive measures triggers creative minds to look for innovative alternative solutions.

Background

"Everyone needs to be engaged"

"We need to be prepared for change"

In order to stimulate climate adaptation at municipal level, the Province of Gelderland initiated Climate Workshops in close collaboration with the Alterra Research Institute of the Wageningen University and Research Centre. In the municipal environment, planning choices are made between issues such as housing, transport, water systems and safety, agriculture, recreation and the natural environment. There is a general understanding of climate change and its uncertainties within the population of the Netherlands. However, the workshops set up in this project aimed to enhance local understanding of the issues in order to start the process of developing climate-proof policies and plans.

Alterra was joined by an independent architectural expert and the Wageningen University to facilitate the workshops. The municipalities also played an important role, providing indispensable information on local characteristics of the area, and designing the 'climate resilient' spatial plans. Disciplines represented at the workshops ranged from (waste) water management, to green space and urban planning and infrastructure, dealing with spatial planning and urban design.

Even though the workshops did not specifically focus on uncertainty, dealing with uncertainty was unavoidable.

Process

"Spread knowledge widely throughout the organization"

An initial workshop was held over 3 days in September 2010 to discuss and create plans to climate-proof specific regions (Fig. 4.31). At this meeting the idea of organising further workshops aimed at individual municipalities was generated. It was felt by the researchers and stakeholders present that if you do not spread climate change related knowledge to everyone in an organisation, then it is wasted. Four of these workshops took place a year later in 2011 with further workshops organised in 2012 and planned for 2013. They bring together many influential individuals round a table to discuss what climate change means for their town. They are usually policy- and decision-makers involved in spatial planning, but aldermen, i.e. senior political representatives of the municipality, have been invited as the ultimate challenge is to engage such politicians.

The workshop process can be roughly divided into the following steps:

- Analysis of the potential **climate change impacts** on a municipal level.
- Assessment of the potential **consequences** of these changes for municipal (spatial) plans.
- **Design sessions to adjust plans** to make them more resilient to a changing climate.
- **Review** of the workshop process, making improvements as necessary and discussion of the process of generating climate-proof spatial plans.

Rather than focussing on changing existing plans the workshops aimed to give the participants a feeling for climate change and adaptation. Actual case studies, relating to water conservation, water nuisance from heavy precipitation, urban heat islands and the robustness and connection of natural areas were used to illustrate the position. Participants attempted to answer the question "how could this plan have been designed to be able to deal with projected climatic changes?" Initially

Fig. 4.31 Workshop in progress

information was presented in a PowerPoint format but as the workshops progressed, various visualisation techniques were developed.

All climate information used during the workshops originated from the Climate Adaptation Atlas (CAA). The adaptation atlas is an ever growing web-portal in which many climate impacts relevant for the Netherlands have been visualised in geospatial maps. It contains maps of projected changes in precipitation, temperature, water nuisance, water safety, droughts, urban-heat-islands and the consequences of these changes for agriculture and nature. It forms a solid foundation of knowledge for the development of adaptation strategies.

Four KNMI scenarios

- W: warm (+2 °C)
- W+: warm + changed air circulation
- G: moderate (+1 °C)
- G+: moderate + changed air circulation

Within the CAA climate uncertainties are addressed by means of scenario analysis, based on the four climate scenarios of the Dutch meteorological office KNMI over four different time steps (2020, 2030, 2050 and 2100). It was important to consider an even number of scenarios to avoid the temptation to focus on a mid-range or average scenario. Precipitation, temperature, water nuisance, water safety, droughts, urban-heat-islands and the consequences of these changes for agriculture, for example in the production of maize, and nature are visualised using the resulting 16 maps (or 17 including the current situation).

Uncertainty Assessment

"Interactive tools allow decision-makers to manipulate the numbers themselves"

The difficulty in presenting such a large number of maps encouraged researchers to seek innovative ways of presenting a broad range of scenario outcomes. How well the information was perceived was subsequently reviewed in detail and the following three different visualisation techniques were experimented with:

- Static visualisation – all maps presented on one page,
- Animated visualisation – an animated presentation displaying a succession of the maps – either over time or across scenarios,
- Interactive visualisation – combination of all maps into one tool, providing a menu to allow a switching between the stacks of images.

Of the three methods presented, the interactive tool, as shown in Fig. 4.32, resulted in the quickest solving of the tasks, giving it the highest score for efficiency. The participants were unanimous in feeling that the interactive tool was the most intuitive. They also liked the ability to continuously compare the different scenarios and time steps with the current scenario.

At the start of the workshops most participants had a good basic knowledge of climate change and its consequences for The Netherlands. However, the extremes and possible range of outcomes were often much greater than expected, and seeing impacts visualised specifically for a municipality was often an eye-opener for them. Practice has shown that single maps are often preferred by decision-makers and are used as predictions rather than being used to explore a range of plausible futures. Also, while policy makers might have a tendency to focus on one of the 'middle of the road' scenario outcomes, scientists often focus on the extremes.

As the design sessions got underway the confrontation with a large range of possible climatic changes and high potential costs of extreme adaptation measures,

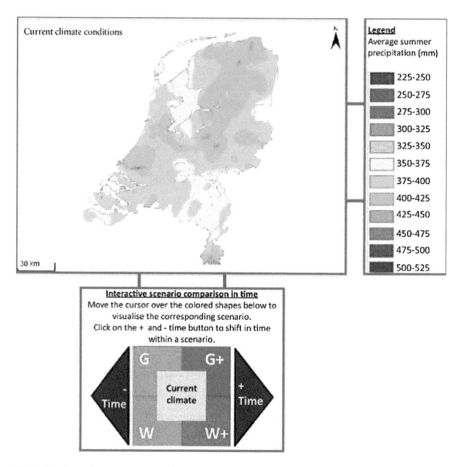

Fig. 4.32 A static representation of the interactive visualisation tool

triggered creative minds to look for innovative, robust measures and to mainstream adaptation measures into other policies. Some examples of this included green roofs as water buffers and insulation, and extra green space in residential areas to increase living comfort.

Effect of Uncertainty on Decision–Making

> *"Decision-makers need to realise they are not 100 %*
> *sure how climate will change"*

The project was primarily designed to communicate the problems of climate change and one of the most significant outcomes was that the project improved the way scenario maps are presented. This is critical to ensure decision-makers fully appreciate the implications of uncertainty in the climate data. Three methods of static visualisation, animated visualisation and interactive visualisation were experimented with. First testing shows that most participants prefer the interactive visualisation as it is the easiest way to handle different information and because of its ability to see patterns in time.

The initial central question of the workshops was 'how can we adapt to climate change?' In the course of the workshops and partly due to the use of a range of scenario outcomes the focus gradually turned towards 'what measures can we take that would allow us to deal with the entire range of possible outcomes?' In one workshop an alderman was looking at houses built in a low, flood-prone part of the region and asked "how could we have been so stupid?" This prompted a rethink of the latest proposal to build on even lower ground, and a realisation of the need to be prepared for change, whatever it might be.

Author: Luuk Masselink

Links for more information:

A general description of the workshops organised at regional level can be found at the website of the national climate programmes of the Netherlands: http://www.klimaatonderzoeknederland.nl/projecten/archief-projecten-nieuws/10657914/Klimaatateliers-COM37.

A report of the Climate Atelier Gelderland on a regional scale can be found at the web portal of the Climate Adaptation Atlas: http://klimaateffectatlas.wur.nl.

The Climate Adaptation Atlas is part of the newly founded foundation Climate Adaptation Services: http://www.climateadaptationservices.com/uk/home

Contact details: luuk.masselink@wur.nl

Chapter 5
Making Adaptation Decisions Under Uncertainty: Lessons from Theory and Practice

Tiago Capela Lourenço, Ana Rovisco, and Annemarie Groot

Key Messages

- Uncertainty can be looked upon from three different points of view:
  - It is possible to deal with uncertainties and act in spite of their existence;
  - It is necessary to reduce uncertainties before making a decision on how to proceed;
  - Uncertainties are considered too large and act either as a barrier to decisions or as a motive to postpone them.

- A clear definition of the adaptation decision objectives and scope is recommended. This will improve communication between decision-makers and those supporting them. Ultimately it will also contribute to enhance the communication between decision-makers and those affected by their decisions (like the public in general or relevant stakeholders).
- The use of multiple methods to deal with and communicate uncertainties is recommended. The correct application of these methods should fit-to-purpose, cover a wide range of uncertainty typologies and aim at providing the widest range of support to different decisions and respective information needs, without compromising clarity.

(continued)

T. Capela Lourenço (✉) • A. Rovisco
Faculty of Sciences, CCIAM (Centre for Climate Change, Impacts, Adaptation and Modelling), University of Lisbon, Ed. C8, Sala 8.5.14,
1749-016 Lisbon, Portugal
e-mail: tcapela@fc.ul.pt; acrovisco@fc.ul.pt

A. Groot
Alterra – Climate Change and Adaptive Land and Water Management,
Wageningen University and Research Centre,
Droevendaalsesteeg 3A, 6708 PB Wageningen, Gelderland, The Netherlands
e-mail: annemarie.groot@wur.nl

(continued)

- Uncertainty can (and should) be communicated in a number of ways:
  - Ensure the involvement of decision-makers and transfer of know-how throughout the development of climate risk and adaptation assessments;
  - Guarantee that messages are clearly communicated and in a language that is common to all stakeholders involved;
  - Promote interactive workshops in order to increase awareness of stakeholders involved;
  - Provide guidance on how to deal with the uncertainties that are present in the outcomes of the decision-making support activity;
  - Use visual depictions of results, including associated uncertainties. For example, the use of interactive tools for visualising scenarios allows stakeholders to handle the data as well as to continuously compare different scenarios and time steps. Other methods of providing visual depictions of results include using confidence scales and score-cards, or recurring to uncertainty typology and ranking of risks according to their likelihood and severity.

- The suggested approaches to decision-making are numerous and should be adjusted to each decision context:
  - Prefer approaches that are robust under a wide range of possible futures, have multiple-benefits and that are low- or no-regret;
  - Prefer options that contribute to enhance resilience and adaptive capacity;
  - Opt for strategies that consider a wide range and variety of options and are able to support adaptive management or learning by doing approaches;
  - Favour options and measures that allow for flexibility.

5.1 Introduction

This chapter synthesises some of the theoretical (scientific) and practical aspects of the preceding chapters, draws key lessons and provides guidance for those involved in supporting and ultimately making adaptation decisions.

A Common Frame of Reference (i.e. common definitions, principles and understandings) for dealing with uncertainties in climate adaptation decision-making is presented and applied to the analysis of the twelve real-life cases presented in this book. A summary of its dimensions and key features is shown in Table 5.1.

This new framework, developed under the scope of the CIRCLE-2 Joint Initiative on Climate Uncertainties,[1] intends to serve as a support to complex climate adaptation decision-making processes that have to deal with uncertainties and still make informed decisions.

---

[1] www.circle-era.eu

Table 5.1 Summary of the Common Frame of Reference dimensions and respective typologies

Dimensions					
	Decision-support				
	To model or not to model?	Top-down or bottom-up?	How certain am I?	Decision-making	Decision-outcomes
Decision-objectives					
Normative/regulatory	Model based	Predictive top-down (optimization or 'science-first')	Statistical uncertainty	Decision made and implementation agreed	Monitoring and evaluating approaches
Strategic/process-oriented	Non-model based	Resilience bottom-up (robustness or 'decision-first')	Scenario uncertainty	Decision delayed	
Operative/action-oriented			Recognised ignorance	Decision not made or not related to adaptation	

Two central questions were addressed using this Common Frame of Reference and were applied to the cases reported in this book:

- How did the approaches used to deal with climate uncertainty influence the adaptation decision-making process?
- Have better informed adaptation decisions been made because uncertainties were conscientiously addressed?

The objective of this chapter is not to provide a simple checklist to be followed when facing uncertainties in a climate adaptation process. Nor does it dare to prescribe a normative 'right' way to make an adaptation decision in the face of climate and non-climate uncertainties.

The purpose here is to inform and guide our readers in navigating a novel, complex and challenging decision-making area, by presenting key lessons and insights from real-life cases were decision-makers and those that support them have already faced and responded to climate adaptation related uncertainty.

As in many other fields, science can inform but in the end decisions are always taken in a 'lonely place'. Despite different cultural contexts, sectors, conditions and ultimately the types of uncertainties that are faced, adaptation decisions are already being made and will continue to be in the foreseeable future. Therefore, the remainder of this chapter presents the reader with the analysis of some hopefully inspiring lessons and approaches that have been followed to support such decisions.

5.2 A New Support Framework for Adaptation Decisions Under Uncertainty

Science-supported decision-making has been the focus of research in multiple scientific and societal challenges (Adger et al. 2013; Ranger et al. 2010; Willows and Connell 2003). Many environmental, economic and societal decision-making processes as well as their underlying knowledge base, tend to be framed from a particular disciplinary perspective (e.g. natural sciences vs. social sciences; basic vs. applied science; technological or economic vs. environmental focus). Climate and climate change adaptation decision-making processes are not a novelty in this regard.

Experience has shown that implementing and communicating climate change impacts and vulnerability assessments in support of practical decision-making is a significant challenge (Tompkins et al. 2010; Adger et al. 2005). Recent literature, mostly concerned with high-end climate change scenarios (e.g. increase of more than 4 °C in global average temperatures) has highlighted some key gaps.

Firstly, the emerging need for innovative strategies and end-user involvement in the development of uncertainty-management methods (Hallegatte 2009). And secondly, the notion that such methods need to be framed within a broader sorting of decision types and systematised into decision support frameworks (Smith et al. 2011).

Climate adaptation decisions, however, are neither taken in isolation from other factors nor are they immune to changes in context specific situations such as culture, economy, politics, resources, institutions, and geography among others (Adger et al. 2008, 2013; Brien et al. 2004).

Adaptation decisions comprise a high level of uniqueness and solutions have often to be determined on a case-by-case approach. Each decision goes through a unique process of development and implementation (Walker et al. 2003). This raises the question of whether it is possible to extract any comparable and valuable lessons from how other decision-makers across the world dealt with uncertainty and ultimately how they came to their adaptation decisions.

Several attempts have been made at capturing and describing the complexity of science-supported climate adaptation decision-making (including policymaking) processes (Hanger et al. 2012; Ranger et al. 2010; Dessai and van der Sluijs 2007; Walker et al. 2003; Willows and Connell 2003).

Nevertheless, practical experience with national and international decision-makers both in Europe as in other parts of the worlds, have shown us how difficult it is to apply such theoretical frameworks into real-life adaptation decisions. Uncertainties in the evidence and in the application of the necessary knowledge base are obviously not the only reason for concern. Yet they rank high when the question at the table is 'how to make an adaptation decision?' or better yet 'how to implement adaptation in practice?'

If positioned in the broader adaptation process context or, for example, as they naturally occur in a risk management cycle, decision-making processes usually encompass some initial framing of the adaptation problem followed by a set of decision-support activities such as research, consulting or policy analysis, the subsequently making of the actual decision and at a later stage the monitoring and evaluation of the decision's outcomes (Hanger et al. 2012; Kwakkel et al. 2011; Ranger et al. 2010; Dessai and van der Sluijs 2007; Walker et al. 2003; Willows and Connell 2003).

There are some key generic features that can be highlighted across these conceptual descriptions of an adaptation decision-making process, namely:

- Their interactive nature;
- The presence of multiple steps (or stages) and feedback mechanisms; and
- Their growing complexity in number and governance of involved agents (both decision-makers and decision-support agents).

Nevertheless, the entry point to these processes is not necessarily always the same and, in practice, the stages in decision-making will not always follow on from one another. It is often necessary to return to previous steps, e.g., to take into account new options only identified after a first round of assessments or appraisal work (Willows and Connell 2003).

Different systems may also need to be assessed differently and pre-exiting conditions may influence the way a decision-maker acts and goes through this cycle. Furthermore, each decision or policy undergoes its own unique process of development and implementation with the involvement of researchers or other kind of analysts potentially taking many different forms (Walker et al. 2003).

Fig. 5.1 A new Common Frame of Reference for science-supported climate adaptation decision-making (This framework has been adapted and modified from Kwakkel et al. (2011), Ranger et al. (2010), Dessai and van der Sluijs (2007), Walker et al. (2003), Willows and Connell (2003) in order to explicitly accommodate the need to deal with uncertainty in the decision-making process)

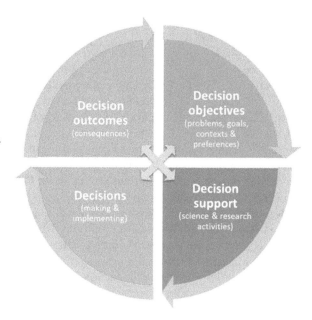

Figure 5.1 describes a simplified Common Frame of Reference to be used in the analysis of a science-supported adaptation decision-making process and as a guiding framework to explore the effect of uncertainties in this sort of decisions. It is based on both academic literature and on the practical experience of dealing with adaptation processes in real-life cases.

It does not intend to be exhaustive but rather to provide a flexible and common approach in understanding how adaptation decision-making under climate change and uncertainty develops, in particular when comparing across different decisions types, decision support methods, and variable geographical, socio-economic and cultural realities.

This Common Frame of Reference is depicted in Fig. 5.1 as a generic cycle involving four inter-connected and complementary dimensions, which can be applied to describe necessary steps in this kind of processes:

- **Decision-Objectives;**
- **Decision-Support;**
- **Decision-Making (and -implementing); and**
- **Decision-Outcomes.**

5.2.1 Decision-Objectives

The entry point to an adaptation decision-making process is often connected with the definition of its objectives. This Decision-Objectives dimension relates to the adaptation problem, as well as to the goals, objectives, values and preferences of the decision-maker and those of the relevant stakeholders.

Choices and decisions will affect the structure and/or performance of the system to which they are applied, so contexts are very important and play a determinant role in this dimension. Although sometimes developed in isolation by decision-makers and their support teams, a decision objective is very often discussed with, or constrained by, stakeholders of all sorts.

Trade-offs between different preferred outcomes that determine the objectives are thus quite important, since adaptation decisions usually have multiple outcomes of interest (Walker et al. 2003).

Within this dimension three common objectives for an adaptation decision can be distinguished, each with its own specificities in terms of uncertainty management:

- **Normative or regulatory**, associated with governance actions that aim to establish a standard or norm;
- **Strategic or process-oriented**, associated with the identification of long-term or overall aims and the necessary setting up of actions and means to achieve them;
- **Operative or action-oriented**, related to the practical actions and steps required to do something, typically to achieve an aim.

5.2.2 Decision-Support

The Decision-Support dimension refers to the set of science, research or other types of activities (like consultancy or policy advice) designed and carried out to support the adaptation decision-makers and the problems being considered.

Scientists, analysts, consultants and other expert advisors are frequently called upon to assess and inform the decision-making process. Often this is the dimension where uncertainties are usually explicitly framed and handled. The uncertainty-management methods and tools described in Chap. 2 and the ones applied in each of the case studies of Chap. 4, are a part of this dimension.

This dimension and the way uncertainties are dealt in it can also be associated to the broader adaptation context as it can usually be seen in, for example, a risk management process cycle. Decision support activities are obviously not exclusive to the adaptation context and are carried out in a variety of policy and decision problems. Lessons can also be learnt there.

In this book we aim exclusively at those activities that are directed at the climate adaptation decision-making and at the way uncertainty is dealt in this particular context. Nevertheless, we do not exclude that this framing of decision typologies and uncertainty management could potentially be useful for other areas of policy and business.

Three generic typologies of relevance to this dimension are detailed below:

To model or not to model?

A common approach to decision support is to create a numerical model of the system, defining its boundaries and structure. It is likely to represent the system's elements and the links, flows and relationships between them (Walker et al. 2003).

In this context, this is termed a model-based decision-support that may or may not be a computer-based model. Non-model decision support (e.g. expert judgement or qualitative assessment) is also commonly employed, in particular when the complexity of the system at hand is too large, or the time availability to coherently model it numerically is too short.

For the sake of simplicity we do not consider 'mental models' as used by experts as part of the model-based support systems (see Lowe and Lorenzoni 2007 and Sect. 2.3.1 of this book).

Models may incorporate different types of uncertainty and because of their common use in this field are often singled out by the public and decision-makers as a primary location of any uncertainty-related problem in the underlying knowledge for adaptation.

These concepts are explored in greater detail in Sect. 2.3.1 of this book.

Top-down or bottom-up?

Another common feature of this dimension is the direction of the approach that is applied to support the decision-making process. In other words, it refers to the direction used by the adaptation assessments or other sort of support activities that are carried out, to the way uncertainties are handled in these and ultimately to the advice they produce.

Such direction is usually defined (Ranger et al. 2010; Dessai and van der Sluijs 2007) as being:

- **Predictive top-down (optimisation or 'science-first')**, emphasising the need to 'foresee' future climate changes and handle the associated uncertainty by categorising, reducing, managing and communicating it. Under this approach the adaptation assessment stages usually follow a linear approach from prediction/projection to decision. They usually begin with projections of climate change, followed by the assessment of potential biophysical impacts and later on by exploring a range of adaptation options;
- **Resilience bottom-up (robustness or 'decision-first')**, accepting uncertainties and unanticipated surprises as being potentially irreducible, and emphasising a 'learning from the past' approach. This approach favours an assessment that usually starts with the adaptation problem at hand (including objectives and constrains), followed by the mapping of available adaptation options, and later evaluating these against projections of climate change.

In reality, mixed approaches are applied in support of adaptation decision-making. This is due to the fact that the choice is not usually between which of the two approaches to use, but rather a need to achieve the best trade-off along a continuous scale that balances between optimisation and robustness (Ranger et al. 2010).

These approaches are explored in greater detail in Sect. 2.5.1 of this book.

How certain am I?

The third feature considered under this dimension is the level of uncertainty that is primarily addressed by the decision-making support activities.

Three levels are distinguished in the literature (e.g. Walker et al. 2003) and, despite the complexity of the concepts, can be analysed in practice:

- **Statistical uncertainty**;
- **Scenario uncertainty**;
- **Recognised ignorance**.

These levels reflect where the uncertainties manifest themselves along a spectrum that progresses from a theoretical full deterministic knowledge of a system ('I'm completely certain of what I know') to an extreme of total ignorance ('I don't even know what I don't know').

The three levels mentioned above lie in between these extremes and represent the most current framing of uncertainty, as it can be regularly applied to practical decision-making support activities (even if not explicitly stated since uncertainties are often not acknowledged).

These levels are explored in greater detail in Sect. 2.3.2 of this book.

5.2.3 Decision-Making

This third dimension of the Common Frame of Reference is related to the actual adaptation decision.

Although there are exceptions, adaptation decisions are usually made in relation to the original problem and objectives, after enough evidence or knowledge has been provided to support an informed action by a decision-maker.

In practice, a decision represents a determination arrived at after consideration, and three results can be associated with an informed adaptation decision-making process under uncertainty:

- **A decision about the adaptation problem is made**, based on the information and evidence provided, and its implementation is agreed and pursued taking into consideration existing uncertainties;
- **A decision is made to delay action regarding the adaptation problem**, until more knowledge is available or the uncertainties associated with the current information or evidence are reduced or differently managed;
- **A decision about the adaptation problem is not made (no-decision)** or a different sort of decision (not related to adaptation or contrary to its objectives) is made and its implementation is agreed and pursued.

These determinations represent, in the context of this book, informed and knowledge-supported decisions normally associated with planned adaptation.

Obviously we cannot have the pretension to map all the contexts where adaptation decisions are made. This means accepting that there can be decisions that are made without explicit external support (such as those related to autonomous adaptation) or yet, that many can be biased by a multitude of factors that have nothing to do with the adaptation problem.

It also means to admit that there will be cases where the information that is provided to a decision-maker may not be the correct one or that science may not always be able to perfectly inform a complex process such as this.

Adaptation decision-making is explored in greater detail in Sects. 2.5 and 2.6 of this book.

5.2.4 Decision-Outcomes

The outcomes of an adaptation decision are difficult to assess and evaluate since some time has to pass (shorter for climate variability and longer for climate change) until the consequences of the decision are visible and can be evaluated. This means that it is also difficult to assess the influence or role played by uncertainty-management methods in shaping up these outcomes.

The monitoring and evaluation (M&E) of adaptation decisions and options has gained recent attention as more and more adaptation decisions are necessary. But adaptation is a relatively recent field of research and especially of decision-making and practice. To date the implementation of adaptation decisions is limited and thus there are not that many outcomes easily available and susceptible of being evaluated. The same applies to the role of uncertainty-management approaches in the shaping of these outcomes.

There has been a recent proliferation of M&E initiatives, guidelines and frameworks. A comprehensive overview of currently available material and tools that can be applied to this dimension is provided by Bours et al. (2013).

Like almost all of the known adaptation examples throughout the world, the real-life cases presented in Chap. 4 have not yet reached this stage, at least from a decisions outcome's evaluation perspective. They can however be the subject of monitoring since they represent adaptation problems that have undergone a decision-making process and that, for better or worse, have seen a given course of action being decided.

Because of the novelty of this dimension there are not many approaches readily available to deal with uncertainties, their contribution to adaptation decisions and its outcomes. Nevertheless, adaptive management approaches have been singled out as being particularly relevant to climate change adaptation and uncertainty management.

Following adaptive management approaches, including monitoring, evaluation and learning (including social learning) that build on growing experience and new knowledge, can also assist in progressive reframing. This is of special relevance being adaptation a continuing and evolving process rather than a single project, decision or initiative (Webb and Beh 2013).

5.3 What Has Practice Shown Us?

In order to better understand how others have dealt with uncertainty in their adaptation decisions and if the processes they followed are transferable, comparability is essential. This section presents some of the key findings extracted from the application of the Common Frame of Reference to the twelve real-life case studies presented in Chap. 4. Table 5.2 presents an overview of key elements, across all cases, for the Decision-Objectives and Decision-Support dimensions.

It allows for a comparative assessment and describes how each situation has dealt with different adaptation objectives and different uncertainty typologies, and how the adaptation decision-making was supported through the use of uncertainty-management and communication methods (see Chap. 2 for more information on the underlying theory).

Each of the case studies is unique in the sense that it tells its own story about policy-makers, decision-makers and scientists who jointly tried to handle the uncertainty inherent to climate change science and move into practice by making informed adaptation decisions.

Table 5.3 further extends this assessment to the third dimension of the Common Frame of Reference, the Decision-Making. In other words, it deals with the adaptation decisions themselves. For each practical case key decisions are presented and a short analysis of how uncertainty played a role in the decision-making process is described.

5.4 Dealing with Uncertainty in Adaptation Decision-Making

Despite the need for 'better' science, this is not in itself a sufficient condition (Tribbia and Moser 2008 and Hanger et al. 2012) for 'better' decisions. These can result from decision-making processes that consider and integrate expert knowledge (Lynch et al. 2008; Dessai et al. 2009), allow for the involvement of relevant stakeholders and that take into account both the climate and non-climate factors representing potential sources of risk and uncertainty (Willows and Connell 2003).

There seems to be a growing consensus that decision-makers are longing for a better integration of existing information rather than more or better information (Tribbia and Moser 2008; Hanger et al. 2012). This must also include the way uncertainty is dealt with along the adaptation decision-making cycle and how uncertainty-management approaches may contribute to a better integration of data sources, processes and knowledge.

Table 5.2 Sorting of the 12 real-life cases (Chap. 4) according to the Common Frame of Reference, dimension further includes the methods used to deal with uncertainty in each case

Cases (Chap. 4)	Decision-Objectives			Decision-Support					
				To model or not to model?		Top-down or bottom-up?		How certain am I?	
	Normative/ regulatory	Strategic/ process-oriented	Operative/ action-oriented	Model based	Non-model based	Predictive top-down	Resilience bottom-up	Statistical	Scenario
Water Supply Management in Portugal (4.2.1)		•	•	•	•	•	•		•
UK Climate Change Risk Assessment (4.2.2)		•		•	•	•		•	•
Water Resources Management in England and Wales (4.2.3)	•			•		•		•	
Water Supply in Hungary (4.2.4)		•		•	•	•			•
Climate Change and Health in The Netherlands (4.2.5)		•			•		•		•
Flood Risk in Ireland (4.2.6)		•		•			•		•
Coastal Flooding and Erosion in South West France (4.2.7)			•	•			•		•
Québec Hydro-Electric Power (4.2.8)		•			•	•			•
Austrian Federal Railways (4.2.9)		•			•		•		•
Dresden Public Transport (4.2.10)		•			•		•		•
Hutt River Flood Management (4.2.11)		•		•	•	•		•	•
Communication of Large Numbers of Climate Scenarios in Dutch Climate Adaptation Workshops (4.2.12)		•			•	•			•
Total	1	10	2	7	9	7	6	3	11

Abbreviations (see Chap. 2 and Key Terms for more detail): *SA* Scenario analysis ('surprise-free'), model ensemble, *BM* Bayesian methods, *NUSAP* NUSAP/Pedigree analysis, *FZ/IP* Fuzzy *EPP* Extended peer review (review by stakeholders), *WC/SS* Wild cards/Surprise scenarios,

for the Decision-Objectives and the Decision-Support dimensions. The Decision-Support

	Methods used to deal with uncertainty												
Recognised ignorance	SA	EE	SENS	MC	PMME	BM	NUSAP	FZ/IP	SI	QA/QC	EPP	WC/SS	Other(s)
	•	•	•						•		•		
	•	•	•			•	•		•	•	•		
				•	•								
	•	•		•				•	•				
•		•							•				
•		•										•	
•		•							•				
	•	•	•		•				•				
	•	•				•			•				
•	•	•	•					•	•			•	•
	•	•		•					•				
	•	•							•				
4	6	9	9	1	4	2	1	2	10	1	2	2	1

EE Expert elicitation, *SENS* Sensitivity analysis, *MC* Monte Carlo, *PMME* Probabilistic multi sets/Imprecise probabilities, *SI* Stakeholder involvement, *QA/QC* Quality assurance/Quality checklists, *Other* Causal and Fuzzy Cognitive Mapping (added by case authors)

Table 5.3 Characterisation and findings for the Decision-making dimension of the real-life cases (presented in Chap. 4)

Cases (Chap. 4)	Decision-Making	
	Have adaptation decisions been made?	What was the influence of uncertainty management in the adaptation decision-making process?
Water Supply Management in Portugal (4.2.1)	*Decision made and implementation agreed*: Establishment of cooperation protocols with external stakeholders. *Decision delayed*: Investments in nanofiltration systems. *Decision not made or not related to adaptation*: Investment decision on prevention measures against forest fires around key water source.	Different initial views of the company's staff were a barrier to adaptation, but the treatment of uncertainties clarified and improved the confidence in the underlying evidence. Started to use multiple-scenarios in the analysis of climate change impacts and vulnerability of water sources. Strategic and operational decisions based on vulnerability assessments that include uncertainty information.
UK Climate Change Risk Assessment (4.2.2)	*Decision made and implementation agreed*: Official use of results and evidence in national and local support of adaptation decision-making (policy and planning).	Priority risks identified with the recognition that uncertainties need to be considered. Incorporation of flexibility into adaptation policies and planning and respective reporting.
Water Resources Management in England and Wales (4.2.3)	*Decision made and implementation agreed*: Development of guidance on the use of probabilistic climate change information in water resources plans.	Acceptance by both the Environment Agency and water companies that planning based on single storylines is a risk in itself. Water companies' willingness to use results originated in the use of multiple models as long as tools remain simple.
Water Supply in Hungary (4.2.4)	*Decision made and implementation agreed*: Establishment of a new system to monitor heavy rains and flash flood in a mountainous area. Installation of a new treatment plan to cope with water quality issues during floods. Shutting down of small water works in low-lying areas. Development of a regional water pipeline to increase water safety. Development of further prospective studies on measures against extreme events.	Despite the use of 3 regional climate models that yielded different results, water companies proposed to accept uncertainty and develop different adaptation measures for the future range of scenarios. Formulation of alternative management measures. Monitoring systems for climate and hydrological parameters considered as essential to deal with uncertainty.

Climate Change and Health in The Netherlands (4.2.5)	*Decision not made or not related to adaptation:* No decisions were made.	Affected how the National Environmental Agency conducts its health assessment for the Dutch Government. Led to the advice that differentiated policy approaches need to be followed according to the characteristics of both health impacts and policy options. Use of uncertainty typologies made uncertainties comparable helping to focus the appropriate policy strategies. Further use of the approach in another agency's study on climate-proofing, (for floods, water availability and urban stress).
Flood Risk in Ireland (4.2.6)	*Decision made and implementation agreed:* To approach decisions using 'softer' techniques in order to ensure robustness and flexibility.	Move from deterministic to robust and flexible approaches on the design of structural flood defences.
Coastal Flooding and Erosion in South West France (4.2.7)	*Decision made and implementation agreed:* To use a 'low regret' approach by restoring sand dunes as flood defences and relocating a road landward, instead of building dykes	Consideration of sea level rise and other drivers beyond climate change in the development of long term coastal defences. Change in local decision-makers' preferences from hard coastal infrastructure (dykes) to 'low regret' solution serving multiple functions (flood and erosion protection, biodiversity, recreation and local economy).
Québec Hydro-Electric Power (4.2.8)	*Decision made and implementation agreed:* To take into account the impacts of climate change in the planning of renovations of hydropower facilities. To review company's position and pursue further in-depth research into cost-benefit adaptation options.	Realisation that more than one climate change scenario is needed to be taken into account. Use of multiple scenarios as varying assumptions for cost-benefit analysis and assessment of the impacts of increased runoff on hydropower assets.
Austrian Federal Railways (4.2.9)	*Decision made and implementation agreed:* To improve railway track drainage in some regions taking into account a range of potential climate changes.	To move towards the climate-proofing of future investments. Realisation that the planning of new infrastructure should not focus on 'optimal' solutions but rather in a range of potential futures. To include climate change into the company's long-term strategy and enhance data collection for trend analysis and monitoring approaches.

(continued)

Table 5.3 (continued)

Cases (Chap. 4)	Decision-Making Have adaptation decisions been made?	What was the influence of uncertainty management in the adaptation decision-making process?
Dresden Public Transport (4.2.10)	<u>Decision delayed</u>: Development of a model for adaptation decision-making and implementation plan for adaptation measures.	A new view on existing strategies and stimulation of actions to address associated uncertainties in relation to adaptation planning. The introduction of new methods and tools for long-term strategic planning including aiming at a future implementation plan for adaptation measures.
Hutt River Flood Management (4.2.11)	<u>Decision made and implementation agreed</u>: Inclusion of the evidence and findings into the review of flood risk management plans.	A shift in thinking from static safety and path dependency to thinking about how to build flexibility into decision making. Sharper consideration of managed retreat as an option for a low-lying area and consideration of a wider range of response options for the Hutt Valley area. Highlighted the need for a continuous consideration of changing climate risks and enabled practitioners to seriously consider complementary measures. Provided local government with a quick response scan framework suitable to be used in the discussion with elected councillors and local urban councils, on the implications of a wide range of options, costs and timings.
Communication of Large Numbers of Climate Scenarios in Dutch Climate Adaptation Workshops (4.2.12)	<u>Decision not made or not related to adaptation</u>: No decisions were made.	Improvement on the way scenarios are presented in the country. Realisation that an interactive way of presenting climate scenarios helps stakeholders to handle a larger number of scenarios and time patterns. Triggered a rethink towards the dealing with the entire range of future climate changes, possible outcomes and alternative adaptive measures.

This has also been argued for by some members of the scientific community who advocate that effective and successful adaptation planning and strategies can be developed and implemented without being significantly limited by the uncertainties present, e.g., in climate projections (Lempert et al. 2004; Hulme and Dessai 2008; Dessai et al. 2009; Lempert and Groves 2010; Walker et al. 2003; Smith et al. 2011).

In fact, Lemos and Rood (2010), go further and state that "there is an uncertainty fallacy", meaning that there seems to be a conviction that for climate projections to be used by decision-makers a reduction in uncertainty is required, which is not always the case.

In this book we looked into these issues from both a theoretical and practical perspective. We had those that need to deal with uncertainty in adaptation decision-making in mind. We believe this group includes not just the decision-makers and practitioners but also all those that support and provide them with the necessary knowledge and evidence.

The following section provides key guidance and recommendations that were extracted from the development and analysis of the twelve practical cases, complemented by the theoretical insights made available to the authors through their research and practice.

5.5 Guidance and Recommendations

Adaptation decisions are a novel area for decision-makers, practitioners and researchers alike. Dealing with uncertainty is a key element for these adaptation decisions. Uncertainty can be looked upon from three different points of view:

- It is possible to deal with uncertainties and act in spite of their existence;
- It is necessary to reduce uncertainties before making a decision on how to proceed;
- Uncertainties are considered too large and act either as a barrier to decisions or as a motive to postpone them.

All three perspectives can be found in practice as seen in Table 5.3 and in Chap. 4 descriptions of the case studies. Since adaptation options may often have associated high costs and major societal implications, the two latter views may be reasonable in particular cases. However, for the majority of adaptation situations including almost all the ones presented here (nine out of twelve cases) the first perspective appears to be the most meaningful and decision-makers do feel that despite existing uncertainties, it is possible to make climate adaptation decisions.

However, there are also cases were decision-makers feel there is a need for reducing uncertainties before investing or deciding upon adaptation measures. In this case, experience shows that (whenever possible) reducing uncertainties in model parameters through a detailed calibration procedure and/or further analysis, or improving their communication, can enhance the confidence on the evidence and make decision-makers more comfortable to act upon the results.

5.5.1 Adaptation Objectives

Setting the scene on an adaptation decision is not an easy task. The analysed cases show the current tendency towards strategic decision objectives (ten out of twelve). This confirms, to some extent, what the literature usually describes as the difficulty in moving adaptation from theory to practice. Strategic decisions are the ones associated with long-term planning and setting of goals. They are related to the development of processes and the setting up of actions (e.g. 'I want an adaptation strategy or plan for my region/city/company').

With some notable exceptions (namely the UK due to its climate change legislative framework), National Adaptation Strategies in European countries (see Chap. 3) or some of the aims proposed by the EU Adaptation Strategy (EC 2013) are examples of such strategic objectives. Instead of asserting norms and regulatory frameworks, these governance pieces seek to map a strategic perspective for decisions and actions to come.

Normative and operational objectives lie on the other extreme of available examples. These may be considered crucial for adaptation but are also harder to find in current practice. For example, in this book only three of the twelve cases describe clearly stated normative or operational objectives, with the latter being found in one single case.

This raises two questions. The first is about the transferability of results from these cases to other regions or countries in terms of uncertainty management and its influence on decisions. The second relates to the cross-analysis of what are the initially described adaptation objectives (see Table 5.2) and what are the actual operational decisions that are made (see Table 5.3).

In the first case, probably only the interested reader can provide an answer. By analysing how uncertainty was dealt in these cases, namely, the 'Water resources management in England and Wales' (normative), the 'Water supply management in Portugal' (strategic and operational) and the 'Coastal flooding and erosion in South West France' (operational), the reader will be able to judge their applicability to a different reality.

The second issue is of a different nature. What practice shows us is that, often, the primary decision-objectives are not clearly stated as being operational, exactly because there is still a lot of novelty in adaptation and because existing uncertainties do not make it easy to move towards real implementation. Nevertheless, operational decisions are being made (see the Hungarian and Austrian cases) even when the original described objective is of a strategic nature.

Uncertainty management and the confidence in the evidence and knowledge provided by support activities seem to play a role here. Changing perspectives about the role of uncertainties in adaptation decisions are a catalyst for operational decision-making even in cases were that was not originally thought of or at least not formulated in such a fashion.

A clear definition of the adaptation decision objectives and scope is recommended. This will improve communication between decision-makers and those supporting them. Ultimately it will also contribute to enhance the

communication between decision-makers and those affected by their decisions (like the public in general or relevant stakeholders).

5.5.2 Decision Support: Uncertainties, Methods and Communication

A multitude of methods and tools are available to deal with uncertainties in support of adaptation decision-making. Table 5.2 presents an overview of methods that were used in each of the case-studies analysed in this book.

All case studies addressed uncertainties related to the climate system and most addressed uncertainties related to both the climate and the human systems.

Reported uncertainties associated to the human system are mainly related with socio-economic developments, demographics and GHG emissions. Uncertainties related to attributes such as ambiguity, including the presence of multiple perceptions about what is known or probable, were not explicitly mentioned. None of the case studies explicitly addressed the (consequences of) relationships between different types of uncertainties.

Three cases reported the use of models as the single approach to support decision-making, while five reported on the use of only non-model based information for this purpose. Four of the cases reported the use of both approaches.

Regarding the direction of the approach followed in support of the decision-making process, six cases reported a top-down/predictive perspective, five a bottom-up/resilience approach and in only one case both were applied.

The correlation between the used of models and the direction of the assessments is important. Only one case used models but reported a bottom-up approach. And none of the cases that reported a top-down approach worked without models.

More than one level of uncertainty was addressed in about half of the cases. Three out of the twelve cases deliberately addressed statistical uncertainty, nine dealt with scenario uncertainty and four with recognised ignorance.

This is in line with our experience since statistical (such as probabilistic data) and recognised ignorance (such as better understanding parts of the system to each the decision is concerned) require not only a larger set of expertise but also considerable amounts of time, not always compatible with the timings decision-makers work with.

Multiple methods are applied to address uncertainty in all case studies. In the large majority of cases these include expert elicitation (ten) and stakeholder involvement (nine). In fact, seven cases applied a combination of both methods, usually in association with other methods.

By large these two methods are the most widely used in uncertainty management at the practical level. Both expert elicitation and stakeholder involvement methods rely heavily on boundary activities between those who support decisions (experts) and those making (decision-makers) or influencing them (stakeholders). This suggests that engagement between such groups is considered critical and it is actively sought out in the support of adaptation decision-making.

In fact only two cases did not report the use of any of these two methods. Interestingly, these represent two of the three cases that applied a 'model only' approach. Yet, even in these cases, meetings with decision-makers (if at an informal level without forming a 'method') to discuss uncertainty and potentially modify perspectives on the issue were mentioned, as in all of the other cases.

Nine of the selected case studies reported the use of sensitivity analysis and less commonly used methods included 'scenario analysis' (six cases) and 'probabilistic multi model ensemble' (four cases). All remaining methods were described either by one or two of the practical case studies.

These results show an interesting landscape. First and foremost a combination of multiple methods is usually applied to address uncertainty. Although it is not possible to correlate the use of methods with the decision objectives, it becomes clear that in order to support complex adaptation decision-making needs, supporting scientists or consultants tend to deploy a large number of methods to deal with uncertainties.

Only three cases used a simple combination of two methods and of those, two applied exclusively expert elicitation together with stakeholder involvement. All other cases used more than four methods in their assessments.

From our experience with these cases, the reason behind the use of such a wide variety of methods is twofold.

Firstly, researchers and others providing support to decision-making recall that, often, decision-makers are not dealing with one single or isolated adaptation decision but with multiple, sometimes even potentially conflicting ones. Furthermore, such decisions are sometimes about different geographical areas. So, in order to fit-to-purpose, the advice on uncertainties that supports multiple adaptation decisions often requires the use of multiple methods, tailored to specific objectives within the assessments.

Secondly, completeness is usually a requirement for decision-making. Having multiple methods involved in the management and communication of uncertainties can enhance the confidence in the information that is provided. This happens because the perception of the decision-maker is changed over time, by getting into contact with these methods, and maybe even being a part of them. Furthermore, methods can be complementary on a given subject and thus provide a more complete assessment of uncertainties.

The use of multiple methods to deal with and communicate uncertainties is recommended. The correct application of these methods should fit-to-purpose, cover a wide range of uncertainty typologies and aim at providing the widest range of support to different decisions and respective information needs, without compromising clarity.

The communication of uncertainties is a key element that needs to be assured not only by those supporting decision-making processes, but also by decision-makers and practitioners themselves, when addressing those affected by their adaptation decisions (general public or specific stakeholders).

Based on both theory and the analysis of the real life practices described in this book, uncertainty can (and should) be communicated in a number of ways:

- **Ensure the involvement of decision-makers and transfer of know-how throughout the development of climate risk and adaptation assessments**;
- **Guarantee that messages are clearly communicated and in a language that is common to all stakeholders involved**;
- **Promote interactive workshops in order to increase awareness of stakeholders involved**;
- **Provide guidance on how to deal with the uncertainties that are present in the outcomes of the decision-making support activity**;
- **Use visual depictions of results, including associated uncertainties. For example, the use of interactive tools for visualising scenarios allows stakeholders to handle the data as well as to continuously compare different scenarios and time steps. Other methods of providing visual depictions of results include using confidence scales and score-cards, or recurring to uncertainty typology and ranking of risks according to their likelihood and severity**.

Although the use of maps and graphs seems to be the most common approach, care should be taken since there is no one-size-fit all approach for the communication of climate change information, regardless of the country or scale of the decision.

5.5.3 Decision-Making and Its Outcomes

The twelve case studies in this book all suggest that as much information as possible should be used so as to avoid poorer adaptation decisions and to better assess the robustness of possible adaptation measures.

However, only two case studies used the information available from the web portals mentioned in Chap. 3, suggesting a need for better integration across scales and dissemination of existing information.

Since climate related uncertainties represent one more issue to consider in the decision-making process of most decision-makers and characterise only a small part of the total risks to be faced, single scenarios should be avoided as the basis of the analysis. All cases support the common notion that no such thing as a "single best scenario" exists for climate change adaptation decision-making, since single scenarios do not represent the full range of possible futures and tend to underestimate extremes.

The analysis of the practical cases has shown that conscientiously addressing uncertainty had an effect on the adaptation decision-making or at best changed attitudes towards climate change adaptation. There is often a clear shift in thinking from a deterministic or 'single optimal solution' approach to adaptation towards a flexible, robust, resilience-oriented and no-regret approach.

The suggested approaches to decision-making are numerous and should be adjusted to each decision context:

- **Prefer approaches that are robust under a wide range of possible futures, have multiple-benefits and that are low- or no-regret;**
- **Prefer options that contribute to enhance resilience and adaptive capacity;**
- **Opt for strategies that consider a wide range and variety of options and are able to support adaptive management or learning by doing approaches;**
- **Favour options and measures that allow for flexibility.**

Because of its novelty, adaptation decisions are yet to be evaluated in regard to their outcomes. Nevertheless, recent literature and several of the cases converge in the notion that monitoring and evaluation methods on one hand and favouring (to the extent possible) adaptive management approaches on the other, can offer a pathway to the future understanding of the consequences of complex adaptation decisions.

5.6 Final Remarks

Adaptation practice is a novel and dynamic field. This is reflected by an as yet limited experience in how climate change uncertainties can be best dealt with in particular situations.

As a consequence, the number of cases in this book can be, to some extent, biased towards the first steps in the development of adaptation policies and strategies (such as the assessment of risk and vulnerability). A significant range of types of decision-making objectives is likely to be underrepresented. The cases that could be included do suggest that often multi-sector and multi-scale decision-processes are covered and indicate that multiple and diverse approaches to inform decisions are applied.

Further research is required to develop methods that evaluate planned and unplanned adaptations and to locate adaptation situations in the landscape of decision-making around risk (Tompkins et al. 2010). Recent literature, mostly related to high-end climate change scenarios (i.e. above 4 °C), has called the attention to some key gaps and requirements of such high-end analysis. It has been suggested that rather than being unable to make decisions under uncertainty, what has been missing is the deployment of innovative decision-making frameworks to deal with uncertainties prompted by climate adaptation assessments (Hallegatte 2009; Smith et al. 2011).

The application of a common frame of reference in the analysis of different types of adaptation decision objectives and of the research approaches used to inform them provides a further step in the understanding of how to design and apply such novel decision-making frameworks (e.g. the role of different information needs vs. different decisions approaches).

Recognizing that site- and culture-specificity of adaptation situations makes generalized conclusions difficult, the work presented in this book aims at advancing the knowledge basis for adaptation decision-making.

By systematically collecting, selecting and analysing concrete examples where science was called upon to support real adaptation decision-making processes using uncertainty management and communication approaches, this book moves us a step closer to the better understanding of two relevant questions.

Firstly, how is science currently dealing with (and communicating) uncertainty in light of existing adaptation decision objectives and needs.

Secondly, what have been the outcomes of such approaches in terms of concrete decisions that were made (or not) and how did the use of different methodologies improve the support to those decision processes ('are better informed adaptation decisions being made?').

The guidance presented here will be subject to further development and enrichment. A growing set of concrete evidence-based adaptation decisions in a variety of situations will provide further stepping-stones towards the improvement of guidance for both decision-makers and researchers involved in climate adaptation decisions.

References

Adger, W.N., N.W. Arnell, and E.L. Tompkins. 2005. Successful adaptation to climate change across scales. *Global Environmental Change* 15(2):77–86. doi:10.1016/j.gloenvcha.2004.12.005.

Adger, W.N., S. Dessai, M. Goulden, M. Hulme, I. Lorenzoni, D.R. Nelson, L.O. Naess, J. Wolf, and A. Wreford. 2008. Are there social limits to adaptation to climate change? *Climatic Change* 93(3–4):335–354. doi:10.1007/s10584-008-9520-z.

Adger, W. Neil, Jon Barnett, Katrina Brown, Nadine Marshall, and Karen O'Brien. 2013. Cultural dimensions of climate change impacts and adaptation. *Nature Climate Change* 3:112–117.

Bours, Denis, Colleen McGinn, and Patrick Pringle. 2013. *Monitoring & evaluation for climate change adaptation: A synthesis of tools, frameworks and approaches*. SEA Change CoP, Phnom Penh and UKCIP, Oxford.

Dessai, Suraje, and Jeroen Van Der Sluijs. 2007. *Uncertainty and climate change adaptation – A scoping study*. Utrecht University, Copernicus Institute, Utrecht, the Netherlands.

Dessai, Suraje, Mike Hulme, and Robert Lempert. 2009. Climate prediction: A limit to adaptation? In *Adapting to climate change: Thresholds, values, governance*, ed. W.N. Adger, I. Lorenzoni, and K. O'Brien, 64–78. Cambridge: Cambridge University Press.

EC. 2013. *COM(2013) 216, an EU strategy on adaptation to climate change*. Brussels: European Commission.

Hallegatte, Stéphane. 2009. Strategies to adapt to an uncertain climate change. *Global Environmental Change* 19(2):240–247. doi:10.1016/j.gloenvcha.2008.12.003.

Hanger, Susanne, Pfenninger Stefan, Dreyfus Magali, and Patt Anthony. 2012. Knowledge and information needs of adaptation policy-makers: A European study. *Regional Environmental Change* 13(1):91–101. doi:10.1007/s10113-012-0317-2.

Hulme, Mike, and Suraje Dessai. 2008. Ventures should not overstate their aims just to secure funding. *Nature* 453(June):979.

Kwakkel, J., M. Mens, A. de Jong, J. Wardekker, W. Thissen, and J. van der Sluijs. 2011. *Uncertainty terminology*. National Research Programme Knowledge for Climate, the Netherlands.

Lemos, Maria Carmen, and Richard B. Rood. 2010. Climate projections and their impact on policy and practice. *Wiley Interdisciplinary Reviews: Climate Change* 1(5):670–682. doi:10.1002/wcc.71.

Lempert, Robert, Nebojsa Nakicenovic, Daniel Sarewitz, and Michael Schlesinger. 2004. Characterizing climate-change uncertainties for decision-makers. *Climatic Change* 65:1–9.

Lempert, Robert J., and David G. Groves. 2010. Identifying and evaluating robust adaptive policy responses to climate change for water management agencies in the American West. *Technological Forecasting and Social Change* 77(6):960–974. doi:10.1016/j.techfore.2010.04.007.

Lowe, Thomas D., and Irene Lorenzoni. 2007. Danger is all around: Eliciting expert perceptions for managing climate change through a mental models approach. *Global Environmental Change* 17(1):131–146.

Lynch, Amanda H., Lee Tryhorn, and Rebecca Abramson. 2008. Working at the boundary: facilitating interdisciplinarity in climate change adaptation research. *Bulletin of the American Meteorological Society* 89(2):169–179. doi:10.1175/BAMS-89-2-169.

O'Brien, Karen, Siri Eriksen, Ane Schjolden, and Lynn Nygaard. 2004. What's in a word? Conflicting interpretations of vulnerability in climate change research. CICERO Working Paper 2004:04, CICERO, Oslo, Norway.

Ranger, Nicola, Antony Millner, Simon Dietz, Sam Fankhauser, Ana Lopez, and Giovanni Ruta. 2010. *Adaptation in the UK: A decision-making process*. GRI and CCCEP, London.

Smith, Mark S., Lisa Horrocks, Alex Harvey, and Clive Hamilton. 2011. Rethinking adaptation for a 4°C world. *Philosophical Transactions of the Royal Society. Series A, Mathematical, Physical, and, Engineering Sciences* 369(1934):196–216.

Tompkins, Emma L., W. Neil Adger, Emily Boyd, Sophie Nicholson-Cole, Keith Weatherhead, and Nigel Arnell. 2010. Observed adaptation to climate change: UK evidence of transition to a well-adapting society. *Global Environmental Change* 20(4):627–635. doi:10.1016/j.gloenvcha.2010.05.001.

Tribbia, John, and Susanne C. Moser. 2008. More than information: What coastal managers need to plan for climate change. *Environmental Science and Policy* 11(4):315–328. doi:10.1016/j.envsci.2008.01.003.

Walker, W.E., P. Harremoees, J. Rotmans, J.P. van der Sluijs, M.B.A. van Asselt, P. Janssen, and M.P. Krayer von Krauss. 2003. Defining uncertainty: A conceptual basis for uncertainty management in model-based decision support. *Integrated Assessment* 4(1):5–17.

Webb, R., and J. Beh. 2013. *Leading adaptation practices and support strategies for Australia: An international and Australian review of products and tools*, 120. Gold Coast: National Climate Change Adaptation Research Facility.

Willows, Robert, and Richenda Connell. 2003. *Climate adaptation: risk, uncertainty and decision-making*. UKCIP Technical Report, UKCIP, Oxford.

Key Terms

This glossary of key terms was compiled by selecting the most relevant terms from various sources such as the IPCC reports (SREX, SRREN and AR4), the RIVM/MNP Guidance on Uncertainty Assessment and Communication and OECD's Adaptation to Climate Change key terms as well as the Climate-ADAPT, EPA and UKCIP online glossaries.

Adaptation Adjustment in natural or human systems in response to actual or expected climatic stimuli or their effects, which moderates harm or exploits beneficial opportunities. Various types of adaptation can be distinguished, including anticipatory, autonomous and planned adaptation. Examples include raising river or coastal dykes, retreating from coastal areas subject to flooding through sea-level rise, or substituting temperature-appropriate or drought-adapted crops for conventional ones.

Adaptation decision-maker Any decision-maker that has to consider climate change in his/her activities and decisions. It is not restricted to persons whose primary task is to address observed and projected impacts of climate change, and it does not intend to suggest that adaptation to climate change is a stand-alone activity.

Adaptive capacity The ability of a system (e.g., an individual, community, society or an organisation) to adjust to climate change (including climate variability and extremes) to moderate potential damages, to exploit beneficial opportunities, or to cope with the consequences.

Adaptation knowledge base Information that is relevant for adaptation planners. The term has also been referred to as "reliable data on the likely impact of climate change, the associated socio-economic aspects and the costs and benefits of different adaptation options".

Adaptation strategy A broad plan of action that is implemented through policies and measures.

Bayesian Method A method of dealing with uncertainties by which a statistical analysis of an unknown or uncertain quantity is carried out in two steps. First, a prior probability distribution is formulated on the basis of existing knowledge (either by eliciting expert opinion or by using existing data and studies). At this first stage, an element of subjectivity may influence the choice, but in many cases, the prior probability distribution is chosen as neutrally as possible, in order not to influence the final outcome of the analysis. In the second step, newly acquired data are introduced, using a theorem formulated by and named after the British mathematician Bayes (1702–1761), to update the prior distribution into a posterior distribution.

Checklist for Model Quality Assistance A method of dealing with uncertainties used to assist modellers and users of models in the process of quality control. The checklist for model quality assistance addresses all sorts of uncertainties at all locations identified in the uncertainty typology. The focus is mainly on unreliability and ignorance and the different sections of the checklist address the different locations in which uncertainty may be found. There are sections on internal strength which address inputs and model structure, and sections on external strength which address system boundary and socio-political context.

Climate Typically defined as the average weather (or more rigorously a statistical description of the average in terms of the mean and variability) over a period of time, usually 30 years. Average weather most often includes surface variables such as temperature, precipitation, and wind. Climate in a wider sense is the state, including a statistical description, of the climate system.

Climate Change This represents any change in climate over time. More specifically it is a change in the state of the climate that can be identified (e.g. using statistical tests) by changes in the mean and/or the variability of its properties and that persists for an extended period, typically decades or longer. It can be due to natural variability or it can be a result of human activity.This definition differs from that in the United Nations Framework Convention on Climate Change (UNFCCC), which defines 'climate change' as 'a change of climate which is attributed directly or indirectly to human activity that alters the composition of the global atmosphere and which is in addition to natural climate variability observed over comparable time periods'. The UNFCCC thus makes a distinction between 'climate change' which it attributes to human activities altering atmospheric composition, and 'climate variability' which it attributes to natural causes.

Climate Model A quantitative way of representing the interactions of the atmosphere, oceans, land surface, and ice. The models are numerical representations of the climate system, based on the physical, chemical, and biological properties of its components, their interactions, and feedback processes. They account for all or some of its known properties. Models can be relatively simple or quite comprehensive. They are applied as a research tool and for operational purposes to study and simulate the climate, and include monthly, seasonal, and interannual climate predictions.

Climate Change Impact, Vulnerability and Risk Assessment This refers broadly to any assessment that systematically assesses the potential environmental, social and/or economic impacts of anticipated climate change.

Climate System This is highly complex and is defined by the dynamics and interactions of five major components: atmosphere, hydrosphere, cryosphere, land surface, and biosphere. Climate system dynamics are driven by both internal and external factors, such as volcanic eruptions, solar variations, or human-induced modifications to the planetary radiative balance. Examples include anthropogenic emissions of greenhouse gases and/or land-use changes.

Climate Variability Variations in the mean state and other statistics (such as standard deviations, the occurrence of extremes, etc.) of the climate on all spatial and temporal scales beyond that of individual weather events. Variability may be due to natural internal processes within the climate system (internal variability), or to variations in natural or anthropogenic external factors (external variability).

Critical Review of Assumptions A method of dealing with uncertainties which enables systematic identification and prioritisation of critical assumptions in (chains of linked) models. It provides a framework for the critical appraisal of model assumptions and typically addresses value-ladenness of choices. The method basically includes all locations that contain implicit or explicit assumptions

Environmental Assessment A procedure that ensures that the environmental implications of decisions are taken into account before a decision is taken. Their purpose is to ensure that programmes and projects likely to have significant effects on the environment are assessed prior to their approval or authorisation. Consultation with the public is a key feature of environmental assessment procedures.

Error Propagation Equations ("Tier 1") Assessment of how the quantified uncertainties in model inputs propagate in model calculations to produce an uncertainty range. This method addresses statistical uncertainty (inexactness) in inputs and parameters and estimates its propagation in simple calculations. It does not treat knowledge uncertainty separately from variability related uncertainty and provides no insight into the quality of the knowledge base or in issues of value loading.

Evidence Based Decision-Making A process for making decisions about a program, practice, or policy that is grounded in the best available research evidence and informed by experiential evidence from the field and relevant contextual evidence.

Expert Elicitation/Expert Judgment A structured process to elicit subjective judgements from experts. It is widely used in quantitative risk analysis to quantify uncertainties in cases where there is no or too few direct empirical data available. In principle, expert elicitation techniques can be tailored and used to elicit and encode subjective expert judgements on any sort of uncertainty at any location identified in the uncertainty typology.

Extended Peer Review (review by stakeholders) Participants in the quality assurance processes of knowledge production and assessment including all stakeholders engaged in the management of the problem at hand. Typically used when facts are uncertain, values in dispute, stakes are high and decisions urgent. It is appropriate when either systems uncertainties or decision stakes are high.

Flexibility In the climate adaptation context, flexibility means the ability to review and adjust strategies as climate change impacts occur through follow-up mechanisms, periodic review and revision of decisions to incorporate new information or data. It allowsdecision-making to be tailored to changing and realistic conditions.

Frequentist Probability This approach repeats a physical process an extremely large number of times ("trials") and then examines the fraction of times that the outcome of interest occurs.

Fuzzy Cognitive Mapping Validated quantitative models of physical, chemical, and biological processes are the best way to assess and project impacts; however, time, data, and model limitations often make these approaches impractical. An alternative is to encode expert knowledge of interactions between system components in a fuzzy cognitive map, which then translates that subjective, qualitative information into predictions of the effects of management on system.

Fuzzy Set/Imprecise Probabilities Options that cannot be expressed as numbers because they are linguistic descriptions of fuzzy perceptions of probabilities (e.g., not very high, quite unlikely, about 0.8, etc.). Such options cannot be assessed through the use of standard probability theory.

General Circulation Model (also known as Global Climate Model or GCM) More commonly known as global climate models, general circulation models are global, three-dimensional computer models of the climate system which can be used to simulate human-induced climate change. GCMs are highly complex and are widely applied for weather forecasting, understanding the climate, and projecting climate change. *See also Climate Model.*

Greenhouse Effect The process by which the absorption of infrared radiation by the atmosphere warms the Earth. The greenhouse effect may refer either to the natural greenhouse effect, due to naturally occurring greenhouse gases, or to the enhanced (anthropogenic) greenhouse effect, which results from gases emitted as a result of human activities.

Greenhouse Gases Those gaseous constituents of the atmosphere, both natural and anthropogenic, that absorb and emit radiation at specific wavelengths within the spectrum of thermal infrared radiation emitted by the Earth's surface, the atmosphere itself, and by clouds. The properties of these gases cause the greenhouse effect.

Water vapour (H_2O), carbon dioxide (CO_2), nitrous oxide (N_2O), methane (CH_4) and ozone (O_3) are the primary greenhouse gases in the Earth's atmosphere. Moreover, there are a number of entirely human-made greenhouse gases in the atmosphere, such as the halocarbons and other chlorine- and bromine-containing substances, dealt with under the Montreal Protocol. Beside CO_2, N_2O and CH_4, the Kyoto Protocol deals with the greenhouse gases sulphur hexafluoride (SF_6), hydrofluorocarbons (HFCs) and perfluorocarbons (PFCs).

Growth Communities Communities that develop, implement, and manage strategies, policies, and programmes with the ultimate purpose of stimulating community economic growth.

Impact Assessment The process of identifying the future consequences of a current or proposed action.

Maladaptation Any changes in natural or human systems that inadvertently increase vulnerability to climatic stimuli; an adaptation that does not succeed in reducing vulnerability but actually increases it.

Key Terms 167

Mitigation A human intervention with the goal of reducing greenhouse gas emissions and/or enhancing sinks.

Monte Carlo Analysis ("Tier 2") A statistical technique for stochastic model-calculations and analysis of error propagation in calculations. The goal of Monte Carlo analysis is to trace the structure of the distributions of model output that result from specified uncertainty distributions of model inputs and model parameters. This method typically addresses statistical uncertainty (stochastic inexactness) in inputs and parameters.

National Adaptation Action Plan These provide guidance on specific national adaptation actions that are being planned. *See also National Adaptation Strategies.*

National Adaptation Strategy (NAS) A broad policy document that outlines the direction of action in which a country intends to move in order to adapt to climate change. While an NAS shows some political commitment towards climate change adaptation, it does not always imply that adaptation activities are taking place.

No-regret Approach An approach that would generate net social and/or economic benefits irrespective of whether or not anthropogenic climate change occurs.

Numeral Unit Spread Assessment Pedigree (NUSAP) (also known as NUSAP/ Pedigree Analysis) A notational system which aims to provide an analysis and diagnosis of uncertainty in science for policy. It captures both quantitative and qualitative dimensions of uncertainty and displays these in a standardised and self-explanatory way. It promotes criticism by clients and users of all sorts, expert and lay, and will thereby support extended peer review processes.

The different qualifiers in the NUSAP system address different sorts of uncertainty. The Spread qualifier addresses statistical uncertainty (inexactness) in quantities, typically in input data and parameters. The Assessment qualifier typically addresses unreliability. The Pedigree criterion further qualifies the knowledge base, exploring the border with ignorance by providing detailed insights in specific weaknesses in the knowledge base that underpins a given quantity.

Percentile A percentile is a value on a scale of 1–100 determined by the percentage of the values in the dataset that are smaller than that value. The percentile is often used to estimate the extremes of a distribution. For example, the 90th (10th) percentile may be used to refer to the threshold for the upper (lower) extremes.

Pluralistic framework of Integrated uncertainty Management and risk Analysis (PRIMA) The guiding principle is that uncertainty legitimises different perspectives and that as a consequence uncertainty management should consider different perspectives. PRIMA is especially suited for uncertainties, which can be interpreted differently from normative standpoints. In practice this usually means that PRIMA is useful for scenario uncertainties and recognised ignorance. The main PRIMA technique of perspective-based multiple model routes, involves both model, input and parameter uncertainties and to a lesser extent the context.

Precautionary measure A precautionary measure is an action taken to avoid a dangerous or undesirable event.

Probabilistic Multi Model Ensemble An ensemble of various climate projection models with model weights being inversely proportional to the random errors in the forecast probability associated with the standard error of the ensemble mean.

Probability Density Function (PDF) The probability density function of a continuous random variable represents the probability that an infinitely small variable interval will fall at a given value. This can be integrated to obtain the probability that the random variable takes a value in a given interval. For example, the probability that a temperature anomaly defined in a particular way is greater than zero is obtained from its PDF by integrating the PDF over all possible temperature anomalies greater than zero.

Quality Assurance/Quality Checklists A process (or set of processes) of enforcing quality control standards by applying planned, systematic activities to examine and improve quality of input, output, and production processes. It examines and controls the formal and systematic use of testing to measure the achievements of specified standards and recommendations.

Regional Climate Model A climate model of higher resolution than a global climate model. It can be nested within a global model to provide more detailed simulations for a particular geographical region (e.g. continent).

Resilience The ability of a social or natural system to absorb disturbances while retaining the same basic structure and ways of functioning; the capacity for self-organisation and the capacity to adapt to stress and change.

Risk-based Decision-making Frameworks Over-arching framing for the assessment and management of risks posed by external and internal drivers on a system of interest for the purpose of identifying potential response options.

Robust decision-making Robust decision-making (RDM) is an iterative decision analytic framework that helps identify potential robust strategies, characterize the vulnerabilities of such strategies, and evaluate the tradeoffs among them. RDM focuses on informing decisions under conditions of what is called 'deep uncertainty,' that is, conditions where the parties to a decision do not know or do not agree on the system model(s) relating actions to consequences or the prior probability distributions for the key input parameters to those model(s).

Robustness The ability of a system to continue to perform satisfactorily under load.

Scenario Analysis (also known as "surprise-free") A method that tries to describe logical and internally consistent sequences of events to explore how the future might, could or should evolve from the past and present. The future is inherently uncertain. Through scenario analysis, different alternative futures can be explored and thus uncertainties addressed. As such, scenario analysis is also a tool to deal explicitly with different assumptions about the future.

Scenario Analysis typically addresses ignorance, value-ladenness of choices (assumptions) and "what-if" questions (scenario uncertainty) with regard to both the context of the (environmental) system considered in the assessment and assumptions about the environmental processes involved. Furthermore Scenario Analysis addresses ignorance, value-ladenness of choices and scenario uncertainty associated with input data and driving forces used in models.

Sensitivity Analysis The study of how the uncertainty in the output of a model (numerical or otherwise) can be apportioned to different sources of uncertainty in the model input. Sensitivity analysis typically addresses statistical uncertainty (inexactness) in inputs and parameters. It is also possible to use this technique to analyse sensitivity to changes in model structure. However, it does not treat knowledge uncertainty separately from variability related uncertainty, and provides no insight into the quality of the knowledge base or in issues of value-loading.

Special Report on Emissions Scenarios (SRES) A report by the Intergovernmental Panel on Climate Change (IPCC) that was published in 2000. The GHG emissions scenarios described in the report have been used to make projections of possible future climate change. These scenarios are often called SRES scenarios.

Stakeholder A person or an organisation that has a legitimate interest in a project or entity, or would be affected by a particular action or policy.

Sustainable Adaptation Adaptation responses that are consistent with and contribute to sustainable development objectives.

UKCP09/UKCIP02 projections The UK Climate Projections which provide climate information designed to help those needing to plan how they will adapt to a changing climate. The climate projections in UKCP09 supersede the scenarios from UKCIP02 (http://ukclimateprojections.defra.gov.uk/21678).

Uncertainty An expression of the degree to which a value (e.g. the future state of the climate system) or relationship is unknown. Uncertainty can result from lack of information or from disagreement about what is known or even knowable. It may have many types of sources, from quantifiable errors in the data to ambiguously defined concepts or terminology, or uncertain projections of human behaviour. Uncertainty can therefore be represented by quantitative measures, for example, a range of values calculated by various models, or by qualitative statements, for example, reflecting the judgement of a team of experts.

Weather The state of the atmosphere with regard to temperature, cloudiness, rainfall, wind, and other meteorological conditions.

Wild Cards/Surprise Scenarios Not sufficiently known risks or opportunities: new futures, new trends, concepts or perceptions.

References

Climate-Adapt. 2012. European climate adaptation platform. http://climate-adapt.eea.europa.eu/glossary. Accessed 27 June 2013.
EC. 2009. White paper – Adapting to climate change: Towards a European framework for action. Available at: http://eur-lex.europa.eu/LexUriServ/LexUriServ.do?uri=CELEX:52009DC0147:en:NOT.
EC. 2013. European Commission. http://ec.europa.eu/ipg/quality_control. Accessed 30 July 2013.
EC. 2013. Adapting to climate change: Towards a European framework for action. Available at: http://eur-lex.europa.eu/LexUriServ/LexUriServ.do?uri=CELEX:52009DC0147:en:NOT.
EPA. 2013. United States Environmental Protection Agency. http://www.epa.gov/climatechange/glossary.html. Accessed 27 June 2013.
Hobbs, Benjamin F., Stuart A. Ludsin, Roger L. Knight, Phil A. Ryan, Johann Biberhofer, and Jan J.H. Ciborowski. 2002. Fuzzy cognitive mapping as a tool to define management objectives for complex ecosystems. *Ecological Applications* 12:1548–1565. http://dx.doi.org/10.1890/1051-0761(2002)012[1548:FCMAAT]2.0.CO;2.
IAIA. 2013. International Association for Impact Assessment. http://www.iaia.org/(X(1)S(klpupmews4rse5owaxqk15zq))/default.aspx?AspxAutoDetectCookieSupport=1. Accessed 30 July 2013.
IPCC. 2007. *Climate change 2007: Impacts, adaptation and vulnerability*. Intergovernmental Panel on Climate Change (IPCC). http://www.ipcc.ch/publications_and_data/publications_ipcc_fourth_assessment_report_wg2_report_impacts_adaptation_and_vulnerability.htm.
IPCC. 2007. *Climate change 2007: The physical science basis*. Intergovernmental Panel on Climate Change (IPCC). http://www.ipcc.ch/publications_and_data/publications_ipcc_fourth_assessment_report_wg1_report_the_physical_science_basis.htm.
IPCC. 2012. Glossary of terms. In: *Managing the risks of extreme events and disasters to advance climate change adaptation*, ed. C.B. Field, V. Barros, T.F. Stocker, D. Qin, D.J. Dokken, K.L. Ebi, M.D. Mastrandrea, K.J. Mach, G.-K. Plattner, S.K. Allen, M. Tignor, and P.M. Midgley. A special report of Working Groups I and II of the Intergovernmental Panel on Climate Change (IPCC), 555–564. Cambridge, UK/New York: Cambridge University Press.
OECD. 2006. Adaptation to climate change: Key terms. COM/ENV/EPOC/IEA/SLT(2006). http://www.oecd.org/environment/cc/36736773.pdf.
Porter, Douglas. R. 2008. *Managing growth in America's communities*. Washington, DC: Island Press.
Steinmüller, K. 2004. Wild cards – What makes them important. http://www.steinmuller.de/media/pdf/WC_GFF.pdf. Accessed 30 July 2013.
UKCIP. 2012. UKCIP. http://www.ukcip.org.uk/glossary. Accessed 27 June 2013.

Van der Sluijs, Jeroen, J.S. Risbey, P. Kloprogge, J.R. Ravetz, S.O. Funtowicz, S. Corral Quintana, Â. Guimaraẽs Pereira, B. De Marchi, A.C. Petersen, P.H.M. Janssen, R. Hoppe, and S.W.F. Huijs. 2003. RIVM/MNP guidance for uncertainty assessment and communication: Detailed guidance. http://www.nusap.net/guidance.

Verbruggen, A., W. Moomaw, and J. Nyboer. 2011. Annex I: Glossary, acronyms, chemical symbols and prefixes. In *IPCC special report on renewable energy sources and climate change mitigation*, ed. O. Edenhofer, R. PichsMadruga, Y. Sokona, K. Seyboth, P. Matschoss, S. Kadner, T. Zwickel, P. Eickemeier, G. Hansen, S. Schlömer, and C. von Stechow. Cambridge, UK/New York: Cambridge University Press.

VOSE Software. 2007. Vose software risk software specialist. http://www.vosesoftware.com/ModelRiskHelp/index.htm#Probability_theory_and_statistics/The_basics/The_definition_of_probability.htm. Accessed 30 July 2013.

WMO. 2013. WMO Lead Centre for Long-range Forecast Multi-model Ensemble. http://www.wmolc.org/multi_model/pMME.php. Accessed 30 July 2013.

Zadeh, L.A. 2002. Toward a perception-based theory of probabilistic reasoning with imprecise probabilities. *Journal of Statistical Planning and Inference* 105:233–264.

Book Reviewers

Please note that reviewers are listed in alphabetical order and in accordance to their involvement in the present publication's chapters.

Chapter 1 Reviewers

Hans-Martin Füssel
Air and Climate Change Programme, European Environment Agency, Kongens Nytorv 6, 1050 Copenhagen K, Denmark
e-mail: martin.fuessel@eea.europa.eu

Annemarie Groot
Alterra – Climate Change and Adaptive Land and Water Management, Wageningen University and Research Centre, Droevendaalsesteeg 3A, 6708 PB Wageningen, The Netherlands
e-mail: annemarie.groot@wur.nl

Tiago Capela Lourenço
University of Lisbon, Faculty of Sciences, CCIAM (Centre for Climate Change, Impacts, Adaptation and Modelling), Ed. C8, Sala 8.5.14, 1749-016 Lisbon, Portugal
e-mail: tcapela@fc.ul.pt

Ana Rovisco
University of Lisbon, Faculty of Sciences, CCIAM (Centre for Climate Change, Impacts, Adaptation and Modelling), Ed. C8, Sala 8.5.14, 1749-016 Lisbon, Portugal
e-mail: acrovisco@fc.ul.pt

Leendert van Bree
Department of Spatial Planning and Quality of Living, PBL Netherlands Environmental Assessment Agency, Oranjebuitensingel 6, 2511 VE The Hague, The Netherlands
e-mail: leendert.vanbree@pbl.nl

Chapter 2 Reviewers

Stefano Caserini
Politecnico di Milano, DIIAR – Sez. Ambientale, Via Golgi 39, 20133 Milano, Italy
e-mail: stefano.caserini@polimi.it

Manuel Gottschick
University of Hamburg, Competence Centrum Sustainable University and Research Center of Biotechnology, Society and Environment Ohnhorststr. 18, 22609 Hamburg, Germany
e-mail: manuel.gottschick@uni-hamburg.de

Anne Knol
National Institute for Public Health and the Environment (RIVM), P.O. Box 1, 3720 BA Bilthoven, The Netherlands
e-mail: anne.knol@rivm.nl

Birgit Kuna
German Aerospace Center (DLR), Linder Höhe, 51147 Cologne, Germany
e-mail: birgit.kuna@dlr.de

Alessandra Laghi
AIAT (Association of Environmental and Planning Engineers), Indica Srl, Via Montebello, 10-44121 Ferrara, Italy
e-mail: alessandra.laghi@libero.it

Jaroslav Mysiak
FEEM (Fondazione Eni Enrico Mattei), Isola di San Giorgio Maggiore, 30124 Venice, Italy
e-mail: jaroslav.mysiak@feem.it

Richard Pagett
Huntersbrook House, Hoggs Lane, Purton, Wiltshire SN5 4HQ, United Kingdom
e-mail: secure@RichardPagett.com

James Risbey
CSIRO (Commonwealth Scientific and Industrial Research Organisation) Marine and Atmospheric Research, GPO Box 1538, Hobart, Tas. 7001, Australia
e-mail: james.risbey@csiro.au

Rob Schoonman
Ministry of Infrastructure and the Environment, PO Box 20951, 2500 EX
The Hague, The Netherlands
e-mail: rob.schoonman@minienm.nl

Liesbeth van Holten
Provincie Utrecht, Department of Physical Environment, Province of Utrecht, P.O.
Box 80300, 3508 TH Utrecht, The Netherlands
e-mail: liesbeth.van.holten@provincie-utrecht.nl

Chapter 3 Reviewers

Malgorzata Bednarek
Chief Inspectorate for Environmental Protection, ul. Wawelska 52/54, 00 922
Warsaw, Poland
e-mail: m.bednarek@gios.gov.pl

Sergio Castellari
CMCC (Centro Euro-Mediterraneo sui Cambiamenti Climatici), via Augusto
Imperatore, 16, 73100 Lecce, Italy
e-mail: sergio.castellari@bo.ingv.it

Maria Manez Costa
Department of Economics and Policy, Helmholtz Center Geesthacht, Climate
Service Center, Fischertwiete 1, 20095 Hamburg
e-mail: maria.manez@hzg.de

Louise Grøndahl
Danish Nature Agency, Haraldsgade 53, 2100 Copenhagen, Denmark
e-mail: logro@nst.dk

Susanne Hanger
IIASA (International Institute for Applied Systems Analysis), Schloßplatz 1, 2361
Laxenburg, Austria
e-mail: hanger@iiasa.ac.at

Jana Kontrošová
Ministry of the Environment, Vršovická 65, Praha 10, Czech Republic
e-mail: jana.kontrosova@mzp.cz

Markus Leitner
EAA (Environment Agency Austria), Spittelauer Lände 5, 1090 Vienna, Austria
e-mail: markus.leitner@umweltbundesamt.at

Susanne Lorenz
University of Leeds, Leeds LS2 9JT, United Kingdom
e-mail: ee08sl@leeds.ac.uk

Jaroslav Mysiak
FEEM (Fondazione Eni Enrico Mattei), Isola di San Giorgio Maggiore, 30124 Venice, Italy
e-mail: jaroslav.mysiak@feem.it

Carin Nilsson
Centre for Environmental and Climate Research, Lund University, Sölvegatan 37, S-223 62 Lund, Sweden
e-mail: carin.nilsson@cec.lu.se

Bertrand Reysset
ONERC (Observatoire National sur les Effets du Réchauffement Climatique), Grande Arche, Tour Pascal A et B, 92055 La Défense CEDEX, France
e-mail: bertrand.reysset@developpement-durable.gouv.fr

Sara Venturini
CMCC (Centro Euro-Mediterraneo sui Cambiamenti Climatici), via Augusto Imperatore, 16, 73100 Lecce, Italy
e-mail: sara.venturini@cmcc.it

Martina Zoller
FOEN (Federal Office for the Environment), Climate Division, 3003 Bern Switzerland
e-mail: martina.zoller@bafu.admin.ch

Chapter 4 Reviewers

David Avelar
University of Lisbon, Faculty of Sciences, CCIAM (Centre for Climate Change, Impacts, Adaptation and Modelling), Ed. C8, Sala 8.5.14, 1749-016 Lisbon, Portugal
e-mail: dnavelar@fc.ul.pt

Manuel Gottschick
University of Hamburg, FSP BIOGUM (Forschungsswerpunkt Biotechnik, Gesellschaft und Umwelt), AgChange, Ohnhorststr. 18, 22609 Hamburg, Germany
e-mail: manuel.gottschick@uni-hamburg.de

Markus Leitner
EAA (Environment Agency Austria), Spittelauer Lände 5, 1090 Vienna, Austria
e-mail: markus.leitner@umweltbundesamt.at

Marjolein Pijnappels
Knowledge for Climate Research Programme, Postbus 85337, 3508 AH Utrecht, The Netherlands
e-mail: m.pijnappels@programmabureauklimaat.nl

Chapter 5 Reviewers

Hans-Martin Füssel
Air and Climate Change Programme, European Environment Agency, Kongens Nytorv 6, 1050 Copenhagen K, Denmark
e-mail: martin.fuessel@eea.europa.eu

Carin Nilsson
SMHI (Swedish Meteorological and Hydrological Institute), Folkborgsvägen 17, SE-601 76 Norrköping, Sweden; and Centre for Environmental and Climate Research, Lund University, Sölvegatan 37, S-223 62 Lund, Sweden
e-mail: carin.nilsson@cec.lu.se

Roger B. Street
UKCIP, Environmental Change Institute, Oxford University, South Parks Road, Oxford OX1 3QY, UK
e-mail: roger.street@ukcip.org.uk

Leendert van Bree
Department of Spatial Planning and Quality of Living, PBL Netherlands Environmental Assessment Agency, Oranjebuitensingel 6, 2511 VE The Hague, The Netherlands
e-mail: leendert.vanbree@pbl.nl

Key Terms Reviewers

Hans-Martin Füssel
Air and Climate Change Programme, European Environment Agency, Kongens Nytorv 6, 1050 Copenhagen K, Denmark
e-mail: martin.fuessel@eea.europa.eu

Annemarie Groot
Alterra – Climate Change and Adaptive Land and Water Management, Wageningen University and Research Centre, Droevendaalsesteeg 3A, 6708 PB Wageningen, The Netherlands
e-mail: annemarie.groot@wur.nl

Tiago C. Lourenço
University of Lisbon, Faculty of Sciences, CCIAM (Centre for Climate Change, Impacts, Adaptation and Modelling), Campo Grande, Ed. C8, Sala 8.5.14, 1749-016 Lisbon, Portugal
e-mail: tcapela@fc.ul.pt

Carin Nilsson
SMHI (Swedish Meteorological and Hydrological Institute), Folkborgsvägen 17, SE-601 76 Norrköping, Sweden; and Centre for Environmental and Climate Research, Lund University, Sölvegatan 37, S-223 62 Lund, Sweden
e-mail: carin.nilsson@cec.lu.se

Ana Rovisco
University of Lisbon, Faculty of Sciences, CCIAM (Centre for Climate Change, Impacts, Adaptation and Modelling), Campo Grande, Ed. C8, Sala 8.5.14, 1749-016 Lisbon, Portugal
e-mail: acrovisco@fc.ul.pt

Roger B. Street
UKCIP, Environmental Change Institute, Oxford University, South Parks Road, Oxford OX1 3QY, UK
e-mail: roger.street@ukcip.org.uk

Leendert van Bree
Department of Spatial Planning and Quality of Living, PBL Netherlands Environmental Assessment Agency, Oranjebuitensingel 6, 2511 VE The Hague, The Netherlands
e-mail: leendert.vanbree@pbl.nl

Index

A
Adaptation, 1–15, 17–38, 41–64, 67–69,
  71–79, 82, 84–96, 98–105, 111, 113,
  116–123, 126, 132–137, 139–161
  action plan, 44, 45
  decision-maker, 46, 61, 145
  decision-making, 4, 9, 17–34, 42, 46, 69,
    76, 140, 142–145, 148–155, 157–160
  decisions, 1–15, 17–38, 46, 56, 58, 61, 69,
    76, 85, 105, 139–160
  knowledge base, 46–62
  pathways, 84
  strategy, 4, 14, 15, 17, 20, 22, 31, 37,
    38, 42–45, 60, 63, 72, 74, 78, 79, 101,
    134, 156
Adaptive capacity, 18, 20, 22, 38, 57, 101,
  140, 160
Advisor, 3, 73, 145

B
Bayesian method, 28, 35, 37, 150
Bottom-up, 29–32, 37, 74, 141, 146, 150, 157
Bours, D., 148
Braun, M., 116

C
Capela Lourenço, T., 67–137, 139–161
Chaumont, D., 116
Checklist for Model Quality Assistance,
Climate, 1, 18, 42, 68, 140
  ADAPT, 42–45
  adaptation, 4, 5, 9, 12, 17–38, 43, 46, 55,
    61, 63, 68, 71, 76, 101, 132, 140–145,
    150, 154, 155, 160, 161

change, 1, 18, 42, 68, 140
change impact, vulnerability and risk
  assessment, 20, 22, 43, 44, 47, 59, 63,
  142, 153
model, 1, 41, 48, 52, 59, 62, 75, 85–88,
  93–95, 104, 114, 115, 119, 123, 129,
  130, 152
projection, 43, 47, 48, 52–56, 58, 61, 62,
  81, 86, 87, 99, 103, 104, 120, 155
scenarios, 1, 12, 29, 31, 36, 42, 43, 48, 49,
  52, 53, 55, 56, 58, 62, 64, 68, 71, 88, 93,
  105, 113, 119, 123, 132–137, 150, 154
system, 18, 23, 26, 28, 47, 82, 99, 112,
  114, 115, 157
variability, 9, 18, 20–22, 28, 31, 148
Common frame of reference, 12, 140–142,
  144, 147, 149, 150, 160
Communicate, 3, 4, 7–11, 15, 16, 36, 54, 72,
  77, 105, 116, 127, 136, 158, 159
Communication, 3, 7, 11, 12, 23, 33, 36, 42,
  44, 47, 50, 54–56, 71, 111, 113, 124,
  132–137, 149, 150, 155–160
Confidence, 83

D
Decision, 1, 18, 42, 68, 140
Decision-maker, 2, 3, 6, 9, 10, 33–36, 38,
  42, 46, 54, 55, 58, 61, 63, 69–72, 77,
  78, 82, 84, 86, 111, 124, 127, 132, 133,
  135–137, 142, 143, 145–149, 153,
  155–159, 161
Decision-making, 2–6, 9, 11, 12, 17–38, 46,
  54, 61, 69, 76–79, 84–86, 90, 94–97,
  101–102, 105–106, 110–111, 115–116,
  120, 124, 128, 130–131, 136, 140–160

T. Capela Lourenço et al. (eds.), *Adapting to an Uncertain Climate:*
Lessons From Practice, DOI 10.1007/978-3-319-04876-5,
© Springer International Publishing Switzerland 2014

Decision-making process, 1, 4–7, 10–12, 21, 24, 69, 84, 140, 142–149, 152–154, 157–160
Decision-objectives, 12, 141, 144–145, 149–151, 156, 158, 160, 161
Decision outcomes, 141, 144, 148–149
Decision-support, 1, 72, 78, 85, 90, 141–147, 149–151, 157–159
Dessai, S., 28, 31, 54
Dimension, 23–25, 27, 30, 31, 107, 140, 141, 144–152
Downscaling, 49, 53, 76, 103, 105, 115

E
Engagement, 3, 7, 12, 72, 157
Environmental assessment, 102
Evidence based decision-making, 161
Experience, 2, 3, 7, 8, 10, 12, 58, 67–137, 142–144, 148, 155, 157, 158, 160
Expert elicitation/expert judgment, 27, 35, 36, 77, 81, 82, 94, 98, 115, 123, 151, 157, 158
Extended peer review (EPR) (review by stakeholders), 28, 35, 77, 150

F
Flexibility, 56, 74, 79, 100
Framework, 9, 11, 19, 20, 28–34, 37, 63, 84, 85, 101, 131, 140, 142–149, 154, 156, 160
Framing, 4, 7, 24, 54, 123, 143, 145, 147, 148
Frequentist probability, 28
Füssel, H.-M., 41–64
Füssler, J., 58
Fuzzy cognitive mapping, 35, 121, 124, 151
Fuzzy set/imprecise probabilities, 28, 35, 95

G
General circulation model/global climate model (GCM), 48, 49, 52, 53, 55, 75, 86, 103, 114, 115, 128, 129
Greenhouse effect, 47
Greenhouse gases (GHG), 19, 23, 32, 47, 114–115, 118, 157
Groot, A., 67–137, 139–161
Growth communities, 127
Guenther, E., 126
Guidance, 2–4, 8, 10, 12, 14, 15, 24, 27, 34, 43, 45–47, 61–63, 88, 90, 118, 131, 140, 152, 155–161
Guide, 5, 22, 61, 64, 86, 90, 142, 148

H
Hanger, S., 44, 45
Hildén, M., 41–64

I
Impact assessment, 20–21, 56, 57, 74, 98
Impact models, 1, 23, 47, 58, 63, 73, 76, 88, 93, 103

K
Klein, R.J.T., 45

L
Lawrence, J., 131
Lemos, M.C., 155
Lessons learned, 2, 3, 7, 8, 14, 15, 96
Level, 1–3, 5, 18, 23–26, 28, 30, 35–38, 42–44, 48, 53–54, 56, 58, 59, 61, 63, 72, 75–77, 81–83, 89, 94, 96, 98–100, 105, 107, 108, 110, 127, 130–133, 137, 143, 147, 153, 157, 158
Location, 5, 21, 24–26, 68, 70, 146
Lopez, A., 90
Lorenz, S., 44–46
Low regret adaptation, 33, 120

M
Maladaptation, 2, 5, 6, 58, 84
Masselink, L., 137
Measure, 20, 25, 26, 31, 32, 38, 54, 69, 73, 75, 76, 78, 81, 84, 92–95, 100, 107, 110, 119, 126, 131, 135, 136, 140, 152, 154, 155, 159, 160
Meyr, J., 126
Mitigation, 19, 20, 32, 126, 129
Model, 1, 18, 42, 68, 140
Monte Carlo analysis ("Tier 2"), 37, 81
Multi-model ensemble, 27, 35, 36, 52, 53, 55, 94, 115, 130, 158
Murphy, C., 106

N
National adaptation action plan (NAAP), 44, 45
National adaptation strategy (NAS), 15, 20, 42–46, 59, 156
Nature, 5–8, 11, 12, 24, 26, 32, 36, 37, 63, 82, 97, 134, 135, 143, 156
Nilsson, C., 1–15

Index

No-regret, 11, 18, 33, 38, 84, 119, 140, 160
  adaptation, 38
  approach, 78, 107, 159
Normative, 23, 24, 141, 142, 145, 150, 156
Numeral Unit Spread Assessment
    Pedigree/pedigree analysis (NUSAP),
    28, 35, 37, 150, 151

O

Operative, 21, 141, 145, 150

P

PDF. *See* Probability density function (PDF)
Percentile, 36, 50, 51, 55, 58, 60, 82, 113,
    114, 129
Pfenninger, S., 44, 45
Policy-maker, 3, 9, 10, 22–24, 31, 33–36, 38,
    46, 63, 69, 132, 135, 149
Practice, 4, 5, 9, 12, 19, 24, 28, 32–34, 69,
    101, 126, 135, 139–161
Practitioner, 3, 10, 38, 67–137, 154, 155, 158
Precaution, 38, 78, 100
Precautionary measure, 38
Prioritise, 23, 33, 34, 75, 79, 83
Probabilistic multi model ensemble, 27, 35,
    36, 94, 130, 158
Probability density function (PDF),
    28, 36, 55, 99
Provider, 3, 4, 54, 56

Q

Qualification of knowledge base, 24, 26–27
Quality assurance/quality checklists, 28, 35, 151

R

RCMs. *See* Regional climate model (RCMs)
Real-life cases, 10, 27, 68, 69, 140, 142, 144,
    148, 150, 152
Recognised ignorance, 26, 35, 37, 96, 141,
    147, 157
Regional climate model (RCMs), 48, 49, 52,
    62, 93–95, 103, 114, 115, 119, 152
Resilience, 22, 29–32, 37, 38, 100, 101, 110,
    140, 141, 146, 150, 157, 159, 160
Reysset, B., 111
Risbey, J., 36
Risk-based decision-making frameworks, 85
Risks assessment, 9–11, 35, 43, 45, 47,
    48, 58–60, 64, 70, 78–85, 99, 100,
    150, 152

Robust, 2, 6, 7, 11, 12, 22, 27, 29, 31, 36–38,
    54, 63, 79, 88, 89, 96, 100–105,
    118–120, 135, 140, 153, 159, 160
  adaptation strategies, 63
  decision-making, 6, 12, 31, 38, 101
Robustness, 22, 27, 29–31, 50, 54, 55, 77, 102,
    103, 133, 141, 146, 147, 153, 159
Rood, R.B., 155
Rovisco, A., 67–137, 139–161
Roy, R., 116
Runhaar, H., 31
Rydell, B., 2

S

Scenario analysis ("surprise-free"),
    27, 35–37, 77, 82, 115, 124,
    130, 132, 150, 158
Scenario uncertainty, 25, 26, 29, 31, 37, 114,
    141, 147, 157
Science-policy interface, 46
Sensitivity analysis, 27, 34–36, 76, 77, 81,
    94, 103–104, 115, 119, 124, 130,
    151, 158
Simonffy, Z., 96
Smith, J.B., 46
Societal, 18, 24, 32, 33, 38, 47, 100, 142, 155
Socio-economic scenario, 1, 56–58, 62,
    74, 75
Sources, 1, 7, 9, 11, 21–23, 25, 30–36, 43–55,
    60–63, 68, 70, 73–75, 77, 78, 84–90,
    92, 93, 95, 96, 103, 111, 112, 118,
    120, 122, 123, 128, 142–143, 149,
    150, 152, 156
Special report on emissions scenarios (SRES),
    48, 52, 56, 57, 75, 86, 93, 129
Stakeholders, 11, 12, 23, 28, 31–33, 35, 36,
    43, 45, 54, 56, 69, 72, 74–80, 82, 86,
    93, 105, 112, 113, 115, 116, 118, 120,
    124, 127, 129, 130, 132, 133, 145,
    149–152, 154, 157–159
Statistical uncertainty, 25, 28, 141, 147, 157
Steinemann, M., 58
Strategic, 60, 72, 77, 78, 85, 90, 96–98, 102,
    108, 111, 116, 121, 122, 126, 128, 141,
    145, 150, 152, 154, 156
Street, R.B., 1–15
Sustainable adaptation, 5

T

Tahy, A., 96
Tang, S., 54
Top-down, 29–32, 37, 74, 141, 146, 150, 157

U

UKCP09/UKCIP02 projections, 48, 49, 52, 54, 55, 58, 64, 81, 82

Uncertainty, 1–15, 17–38, 41–64, 69, 72, 75–79, 82–84, 88–90, 93–96, 98–105, 110, 112, 114, 115, 117, 119–121, 124, 127–130, 132, 133, 135, 136, 139–161
    assessment, 5, 9, 17–38, 56, 76–77, 82, 88–89, 93–95, 98–100, 103–105, 110, 114–115, 119–120, 124, 129–130, 135–136
    management, 142, 145, 148, 149, 152–154, 156, 157, 160

V

Value-ladenness, 17, 24, 27
van Bree, L., 17–38
van der Sluijs, J., 17–38, 101
Visual depiction, 12, 127, 140, 159

W

Walker, W.E., 24
Walton, P., 2
Wardekker, A., 31, 101
Watts, G., 90
Weather, 21, 47, 81, 103, 118, 121–125
Web portal, 10, 41–43, 55, 56, 62–64, 134, 137, 159
Wild cards/surprise scenarios, 28, 31, 35, 37, 104, 124, 150

CPSIA information can be obtained
at www.ICGtesting.com
Printed in the USA
LVHW081036310520
657056LV00002B/40